515
San
Ord.
s

at:

HAY LIBRARY
WESTERN WYOMING COMMUNITY COLLEGE

Ordinary Differential Equations

A Brief Eclectic Tour

© 2002 by

The Mathematical Association of America (Incorporated)

Library of Congress Catalog Card Number 2002101380

ISBN 0-88385-723-5

Printed in the United States of America

Current Printing (last digit):

10 9 8 7 6 5 4 3 2 1

Ordinary Differential Equations

A Brief Eclectic Tour

David A. Sánchez

Texas A&M University

Published and distributed by
The Mathematical Association of America

CLASSROOM RESOURCE MATERIALS

Classroom Resource Materials is intended to provide supplementary classroom material for students—laboratory exercises, projects, historical information, textbooks with unusual approaches for presenting mathematical ideas, career information, etc.

Committee on Publications
Gerald Alexanderson, *Chair*

Zaven A. Karian, *Editor*

Frank Farris	David E. Kullman
Julian Fleron	Millianne Lehmann
Sheldon P. Gordon	William A. Marion
Yvette C. Hester	Stephen B Maurer
William J. Higgins	Edward P. Merkes
Mic Jackson	Judith A. Palagallo
Paul Knopp	Andrew Sterrett, Jr.

101 Careers in Mathematics, edited by Andrew Sterrett

Archimedes: What Did He Do Besides Cry Eureka?, Sherman Stein

Calculus Mysteries and Thrillers, R. Grant Woods

Combinatorics: A Problem Oriented Approach, Daniel A. Marcus

Conjecture and Proof, Miklós Laczkovich

A Course in Mathematical Modeling, Douglas Mooney and Randall Swift

Cryptological Mathematics, Robert Edward Lewand

Elementary Mathematical Models, Dan Kalman

Geometry From Africa: Mathematical and Educational Explorations, Paulus Gerdes

Interdisciplinary Lively Application Projects, edited by Chris Arney

Laboratory Experiences in Group Theory, Ellen Maycock Parker

Learn from the Masters, Frank Swetz, John Fauvel, Otto Bekken, Bengt Johansson, and Victor Katz

Mathematical Modeling in the Environment, Charles Hadlock

Ordinary Differential Equations: A Brief Eclectic Tour, David A. Sánchez

A Primer of Abstract Mathematics, Robert B. Ash

Proofs Without Words, Roger B. Nelsen

Proofs Without Words II, Roger B. Nelsen

A Radical Approach to Real Analysis, David M. Bressoud

She Does Math!, edited by Marla Parker

Solve This: Math Activities for Students and Clubs, James S. Tanton

MAA Service Center
P.O. Box 91112
Washington, DC 20090-1112
1-800-331-1MAA   FAX: 1-301-206-9789

Preface — Read This First!

A Little History

The study of ordinary differential equations is a rich subject, dating back to even before Isaac Newton and his laws of motion, and is one of the principal building blocks to describe a process wherein a quantity (mass, current, population, etc.) exhibits change with respect to time or distance. Prior to the late 1800s, the large part of the study of differential equations was devoted to trying to analytically describe solutions in closed forms or via power series or integral representations. The differential equations were almost lost in the sea of special functions (Bessel, Legendre, hypergeometric, etc.) which were their solutions.

But with the publication of Henri Poincaré's memoir on the three-body problem in celestial mechanics in 1890, and his later three volume work on the topic, the subject took a dramatic turn. Analysis was buttressed with geometry and topology to study the dynamics and stability of ordinary differential equations and systems thereof, whose solutions could only be approximated at best. Thus was created the *qualitative* theory of ordinary differential equations and the more general subject of *dynamical systems*. Unfortunately, up until the last half of the past century, very little of the theory was available to students (with the exception of Russia where a large, influential school flourished). Differential equations texts for undergraduates were largely dull "plug and chug" expositions which gave the lasting impression that the subject was a bag of tricks and time consuming infinite series or numerical approximation techniques. Some faculty still believe this.

But with the advances made in computers and computer graphics, available to students through programs like MAPLE, MATLAB, MATHEMATICA, etc., the subject and the textbooks took on a new vitality. Especially important was the access to very powerful numerical techniques such as RKF45 to approximate solutions to high degrees of accuracy, coupled with very sophisticated graphics to be able to display them. Now *qualitative* theory is linked with *quantitative* theory, making for a very attractive course providing valuable tools and insights. Furthermore, mathematical modeling of physical, biological, or socioeconomic problems using differential equations is now a major component of undergraduate textbooks. The slim, dry tomes of the pre-1960s, filled with arcane substitutions and exercises more intended to assess the reader's integration techniques, have been replaced with 600 plus page volumes crammed with graphics, projects, pictures and computing.

The Author's View

I have taught ordinary differential equations at all audience levels (engineering students, mathematics majors, graduate students), and have written or coauthored three books in the subject, as well as over forty research papers. One of the books was a forerunner of the colossal texts mentioned above, and may have been the first book which blended numerical techniques within the introductory chapters, instead of lumping them together in one largely ignored chapter towards the back of the book.

Several years ago, I reviewed a number of the current textbooks used in the introductory courses (*American Mathematical Monthly*), and the article was intended to ask the question "Why have ordinary differential equation textbooks become so large?" But on later reflection I think the right question would have been whether the analytic and geometric foundations of the subject were blurred or even lost in the exposition. This is a question of real importance for the non-expert teaching the course, and of greater importance to the understanding of the student.

Pick up almost any of the current texts and turn to the first chapter, which usually starts off with a section entitled *Classification of Differential Equations*. This is comprised of examples of linear, nonlinear, first order, second order, et al., explicit and implicit ordinary differential equations, with a few partial differential equations thrown in for contrast, and probably confusion. Then there follows a set of problems with questions like "For the following differential equations identify which are linear, nonlinear, etc. and what are their orders?"

This section is usually followed by a discussion usually named *Solutions or Existence of Solutions*, and the students, who may have a hazy understanding at best of what are the objects under examination, are now presented with functions (which come from who knows where) to substitute and to verify identities with. Or they may be presented the existence and uniqueness theorem for the initial value problem, then asked to verify for some specific equations whether a solution exists, when they are not at all sure of what is a solution, or even what is a differential equation.

This confusion can linger well into the course, but the student may gain partial relief by finding integrating factors, solving characteristic polynomials, and comparing coefficients, still unclear about what's going on. This insecurity may be relieved by turning to the computer and creating countless little arrows waltzing across the screen, with curves pirouetting through them which supposedly describe the demise of some combative species, the collapse of a bridge, or even more thrilling, the onset of chaos. I do not blame the well qualified authors of these books; they have an almost impossible task of combining theory, technology, and applications to make the course more relevant and exciting, and try and dispel the notion that ordinary differential equations is a "bag of tricks." But I think it is worthwhile to occasionally "stop and smell the roses," as they say, instead of galumphing through this rich garden, and that is the purpose of this little book.

A Tour Guide

The most important points are

1. *This is not a textbook, but is instead a collection of approaches and ideas worth considering to gain further insight, of examples or results which bring out or amplify an important topic or behavior, and an occasional suggested problem. Current textbooks have loads of good problems.*

2. *The book can be used in several ways:*

   a. It can serve as a resource or guide in which to browse for the professor who is embarking on teaching an undergraduate course in ordinary differential equations.

   b. It could be used as a supplementary text for students who want a deeper understanding of the subject.

   c. For a keen student it could serve as a textbook if supplemented by some problems. These could be found in other introductory texts or college outlines (e.g. Schaum's), but more challenging would be for the student to develop his or her own exercises!

Consequently, the book is more *conceptual than definitive*, and more *lighthearted than pedagogic*.

There is very little mathematical modeling in the book; that area is well covered by the existing textbooks and computer-based course materials. I am a devotee of population models and nonlinear springs and pendulums, so there may be a few of them, but mainly used to explain an analytic or geometric concept. The reader may have to fill in some calculations, but there is no list of suggested problems, except as they come up in the discussion. Hopefully, after reading a section, the reader will be able to come up with his or her own illuminating problems or lecture points. If that occurs then my goal has been met.

A brief description of each chapter follows; it gives the highlights of each chapter. The order is a standard one and fits most textbooks.

Chapter 1: Solutions

The notion of a solution is developed via familiar notions of a solution of a polynomial equation, then of an implicit equation, from which the leap to a solution of an ordinary differential equation is a natural one. This is followed by a brief introduction to the existence and uniqueness theorem, and the emphasis on its local nature is supported by a brief discussion of continuation—rarely mentioned in introductory texts. The discussion is greatly expanded in Chapter 2.

Chapter 2: First Order Equations

The two key ordinary differential equations of first order are the separable equation and the linear equation. The solution technique for the first is developed very generally, and leads naturally into an introductory discussion of stability of equilibria. The linear equation is discussed as a prelude to the higher dimensional linear system, with the special technique of integrating factors taking a back seat to the more general variation of parameters method. Some other topics mentioned are discontinuous inhomogeneous or forcing terms, and singular perturbations.

The Riccati equation, the author's favorite nonlinear equation, is discussed next because it exemplifies many of the features common to nonlinear equations, and leads into a very elegant discussion of the existence of periodic solutions. This is a topic rarely discussed in elementary books, yet it gives nice insights into the subject of periodic solutions of linear and nonlinear equations. There is an expanded discussion in a later section, which includes a lovely fixed point existence argument.

The chapter concludes with a brief analysis of the question of continuous dependence on initial conditions, often confused with sensitive dependence when discussing chaotic behavior. Finally, we return to the subject of continuation, and the final example gives a nice picture of "how far can we go?"

Chapter 3: Insight Not Numbers

The chapter is not intended as an introduction to numerical methods, but to provide a conceptual platform for someone faced with the often formidable chapters on numerical analysis found in some textbooks. There must be an understanding of the idea that numerical schemes basically chase derivatives, hence the Euler and Improved Euler method are excellent examples. Next, the notion of the error of a numerical approximation must be understood, and this leads nicely into a discussion of what is behind a scheme that controls local error in order to reduce global error. The popular RKF45 scheme is discussed as an example.

Chapter 4: Second Order Equations

In a standard one semester course the time management question moves center stage when one arrives at the analysis of higher order linear equations, and first order linear systems. One is faced with characteristic polynomials, matrices, eigenvalues, eigenvectors, etc., and the presentation can become a mini-course in linear algebra, which takes time away from the beauty and intent of the subject. For the second order equation the author suggests two approaches:

a) Deal directly with the general time-dependent second order equation, developing fundamental pairs of solutions and linear independence via the Wronskian.

b) Develop the theory of the two-dimensional time-dependent linear system, again discussing fundamental pairs of solutions whose linear independence is verified with the Wronskian. Then consider the special case where the linear system represents a second order equation.

The emphasis in both approaches is on fundamental pairs of solutions, and linear independence via the Wronskian test, and not spending a lot of time on linear algebra and/or linear independence per sé.

The discussion continues with a now easily obtained analysis of the constant coefficient second order equation and damping. This is followed by the non-homogeneous equation and variation of parameters method, pursuing the approach used for the first order linear equation, and using the linear system theory, previously developed. The advantage is that one can develop the usual variation of parameters formula, without appealing to the "mysterious" condition needed to solve the resulting set of equations.

The section *Cosines and Convolutions* is a discussion of the inhomogeneous constant coefficient equation where the forcing term is periodic. This leads to a general discussion of the onset of resonance, and for the forced oscillator equation we develop a nice convolution representation of the solution which determines whether resonance will occur for any periodic forcing term. The representation can be developed directly, or from the variation of parameters formula with more effort; Laplace transform theory is not needed.

The final section of the chapter gives some of the author's views on various topics usually covered under the second order linear equation umbrella. Infinite series solutions is another point in the introductory course where one must make decisions on how deeply to proceed. The author has some advice for those who wish to move on but want to develop some understanding. There is no mention of the Laplace transform—any kind of a proper treatment is far beyond the intent of this book, even with the nice computer packages which have eliminated the dreary partial fraction expansions.

Chapter 5: Linear and Nonlinear Systems

Given the material in the previous chapter it is now an easy task to discuss the constant coefficient linear system, for both the homogeneous and inhomogeneous case (the variation of parameters formula). But for the two-dimensional case, the important topic is the notion of the phase plane and this is thoroughly analyzed.

However, the character of almost all the possible configurations around an equilibrium point is more easily analyzed using the two-dimensional representation of the second order scalar equation. The trajectories can be drawn directly and correspond to those in the more general case via possible rotations and dilations. This approach virtually eliminates all discussion of eigenvectors and eigenvalues, which is in keeping with the spirit of the book.

Next is given a general rationale, depending on the per capita growth rate, for the construction of competition and predator prey models of population growth. This makes it easier to explain particular models used in the study of populations, as well as to develop one's own. In Chapter 1 the constant rate harvesting of a population modeled by the logistic equation was discussed, and a nice geometric argument led to the notion of a critical harvest level, beyond which the population perishes. A similar argument occurs for a two population model in which one of the populations is being harvested—this is rarely discussed in elementary books.

There follows a section on conservative systems which have the nice feature that one can analyze their stability using simple graphing techniques without having to do an eigenvalue analysis, and also introduce conservation of energy. The book concludes with a proof, using Gronwall's Lemma, of the Perron/Poincaré result for quasilinear systems, which is the foundation of much of the analysis of stability of equilibria. It is a cornerstone of the study of nonlinear systems, and seems a fitting way to end the book.

Conclusion

I hope you enjoy the book and it gives you some conceptual insights which will assist your teaching and learning the subject of ordinary differential equations. Perhaps the best summary of the philosophy you should adopt in your reading was given by Henri Poincaré:

> In the past an equation was only considered solved when one had expressed the solution with the aid of a finite number of known functions; but this is hardly possible one time in a hundred. What we should always try to do, is to solve the qualitative problem, that is to find the general form of the curve representing the unknown function.

> Henri Poincaré
> King Oscar's Prize, 1889

Contents

1

Solutions

To begin to understand the subject of ordinary differential equations (referred to as ODEs in the interest of brevity) we first need to understand and answer the question "What is a solution?". Perhaps the easiest way is to start the hierarchy of solutions way back in our algebra days and move towards our desired goal.

1 Polynomials

Consider a general polynomial of degree n,

$$P(x) = \sum_{k=0}^{n} a_k x^k$$

and suppose we wish to find a solution of the equation $P(x) = 0$. In this case a solution is a *number* x_0 satisfying $P(x_0) = 0$. The first question to ask is that of *EXISTENCE*—does a solution exist? If the coefficients a_k are real the answer is *MAYBE* if we want real solutions—contrast

$$P(x) = x^2 - 1 \quad \text{and} \quad P(x) = x^2 + 1.$$

But if we extend the domain of the coefficients and solutions to be the complex numbers, then the answer is *YES*, by the Fundamental Theorem of Algebra.

The theorem tells us more—it says that there will be n solutions, not necessarily all distinct, so the question of *UNIQUENESS*, except in the case of $n = 1$ when there is exactly one solution of $P(x) = 0$, is essentially a moot one. The question CAN WE FIND THEM? would suggest algorithms for finding solutions. They exist for $n = 2$ (the quadratic formula), $n = 3$ and $n = 4$, but beyond that the best we can usually do is *numerically* approximate the solutions, i.e., "solve" approximately.

1

2 More General Equations

For equations $F(x) = 0$ where $F\colon R \to R$ or $F\colon R^n \to R$ is a non-polynomial equation, there are no general results regarding existence or uniqueness, except some general fixed point results (Brouwer's Theorem), and for $n = 1$ something simple like

> If F is continuous and $F(x_0) > 0, F(x_1) < 0$, $x_0 < x_1$, then $F(x) = 0$ has at least one solution in the interval $x_0 < x < x_1$.

Equations may have many solutions e.g., $F(x) = \sin x - x/60$, or none, e.g., $F(x) = e^x - x^{1/5}$, and solution techniques involve some approximation scheme like Newton's Method.

Remark. An exercise to show the value of some pen and paper calculation before calling in the number crunching is to find the next to last zero of the first function, where the zeros are numbered consecutively starting with $x_1 = 0$.

3 Implicit Equations

Now consider the equation $F(x, y) = 0$, where for simplicity we can assume that $x, y \in R$, though the case for $x \in R$, $y \in R^n$, is a simple generalization. The question is "Can we solve for y?" that is, can we find a *solution* $y(x)$, a *function* satisfying $F(x, y(x)) = 0$. In some cases the solution can be

> unique: $x^2 - \ln y = 0; y = e^{x^2}$,
> not unique: $-x^2 + y^2 - 4 = 0; y = \sqrt{4 + x^2}, y = -\sqrt{4 + x^2}$,
> defined only on a finite interval: $x^2 + y^2 - 4 = 0$;
>
> $$y = \pm\sqrt{4 - x^2} \text{ only defined for } |x| \le 2,$$
>
> unbounded: $x^2 + \frac{1}{y^2} - 9 = 0; y = \pm\frac{1}{\sqrt{9-x^2}}$ is only defined for $|x| < 3$,
> not explicitly solvable: $F(x, y) = x^2 + 4xy^2 + \sin(\pi y) - 1 = 0$.

What is needed is some kind of an *Existence and Uniqueness Theorem*, and it is provided by the Implicit Function Theorem:

> Given $F(x, y) = 0$ and *initial conditions* $y(x_0) = y_0$ satisfying $F(x_0, y_0) = 0$, suppose that $F(x, y)$ is continuously differentiable in a neighborhood of (x_0, y_0) and $\partial F/\partial y(x_0, y_0) \neq 0$. Then there is a unique solution $y(x)$, defined and continuous in some neighborhood N of (x_0, y_0), satisfying the initial conditions, and $F(x, y(x)) = 0$, for $x \in N$.

The theorem is a *local* existence theorem, inasmuch as the solution may be only defined near (x_0, y_0):

Example. $F(x, y) = x^2 + y^2 - 4 = 0$, and $\frac{\partial F}{\partial y} = 2y \neq 0$ for $y \neq 0$. Let $(x_0, y_0) = (1.8, \sqrt{0.76})$, then the solution $y(x) = \sqrt{4 - x^2}$ is only defined for $-3.8 \le x - 1.8 \le 0.2$, $0 \le y \le 2$.

Furthermore, the solution given by the theorem is *unique*, meaning that there cannot exist two continuous functions which are solutions and whose graphs intersect at (x_0, y_0).

Can we find solutions? Only numerically is the usual answer, using a combination of numerical and implicit plotting techniques. For the example above $F(x, y) = x^2 + 4xy^2 + \sin \pi y - 1 = 0$, $F(1, 0) = 0$, and $\partial F / \partial y (1, 0) = -\pi$, and a plot of the solution (see diagram) shows that the solution $y(x)$ satisfying $y(1) = 0$ is only defined for approximately $0 < x < 1.2$.

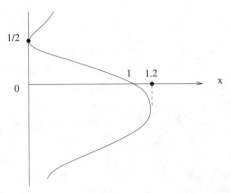

4 Ordinary Differential Equations

We generalize from the previous case $F(x, y) = 0$ and consider an equation of the form $F\left(x, y, \frac{dy}{dx}\right) = 0$, or $F\left(t, x, \frac{dx}{dt}\right) = 0$, depending on the choice of x (space related) or t (time related) as the independent variable. In the previous case the solution was a continuous function—now a *solution* is a *differentiable function* $y(x)$ or $x(t)$ satisfying

$$F\big(x, y(x), y'(x)\big) = 0 \quad \text{or} \quad F\big(t, x(t), \dot{x}(t)\big) = 0.$$

The equations above are *first order ordinary* differential equations. They are

first order—only a first derivative of the solution appears.

ordinary—only the ordinary derivative $y'(x)$ or $\dot{x}(t)$ appears, not any partial derivatives. There is only one independent variable.

Example. $F\left(t, x, y, \frac{\partial y}{\partial t}, \frac{\partial y}{\partial x}\right) = 0$, would be a first order partial differential equation with solution $y(x, t)$, dependent on two variables.

The definition above can be extended to equations or systems like

a) $F\left(x, y, \frac{dy}{dx}, \frac{d^2 y}{dx^2}\right) = 0$, a second order ODE with solution a twice differentiable function $y(x)$,

b) $\left. \begin{array}{l} F_1\left(t, x, y, \frac{dx}{dt}\right) = 0 \\ F_2\left(t, x, y, \frac{dy}{dt}\right) = 0 \end{array} \right\}$.

A two-dimensional system of first order ODEs, whose solution is a pair of scalar differentiable functions $x(t), y(t)$.

These generalizations can be further extended, but understanding is best acquired using the simplest case.

The ODE in its implicit form $F(t, x, \dot{x}) = 0$ is not very useful nor does it occur frequently in the real world, since we usually describe a physical process in the form

(rate of change) = (function of independent and dependent variables, parameters).

Hence, we assume we can solve $F(t, x, \dot{x}) = 0$ for $\dot{x}$ and write the first order ODE as

★ $\dfrac{dx}{dt} = f(t, x)$ (we have chosen t, x as the variables—use x, y if you prefer).

Some examples and their solutions follow; you may wish to substitute the solution $x(t)$ to obtain an equality $\dot{x}(t) = f(t, x(t))$. *Do not* be very concerned about how the solutions were obtained—that comes later. The present object is to understand the notion of a solution.

ODE	Solution
$\dot{x} = 2x$	$x(t) = 3e^{2t}$, or $x(t) = Ce^{2t}, C$ constant
$\dot{x} = 1 + x^2$	$x(t) = \tan t$, or $x(t) = \tan(t - C), C$ constant
$\dot{x} = 3t^2 + 4$	$x(t) = t^3 + 4t + C, C$ any constant
$\dot{x} = -x + 4t$	$x(t) = Ce^{-t} + 4t - 4, C$ any constant
$\dot{x} = rx\left(1 - \frac{x}{K}\right)$	$x(t) \equiv K$ or $x(t) \equiv 0$ or
$r, K > 0$	$x(t) = \frac{K}{1 - (1 - \frac{K}{x_0})e^{-rt}}, x(0) = x_0 \neq 0$

Suppose $x(t)$ is a solution of ★, then we can write $\dot{x}(t) = f(t, x(t))$ and if we integrate with respect to t this implies

$$x(t) = \int^t f(s, x(s))ds + C = F(t, C)$$

where $F(t, C)$ represents the indefinite integral, and C is an appropriate constant of integration. So a general solution will be of the form $x(t, C)$ and we obtain the notion of a family of solutions parameterized by C.

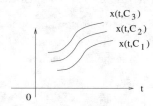

But we can go further; suppose we are given the *initial condition* $x(t_0) = x_0$. Then we want the constant C to be chosen accordingly, and can write the formal expression for the solution

$$x(t) = x_0 + \int_{t_0}^t f(s, x(s))ds.$$

It is easily verified that $x(t_0) = x_0$ and $\dot{x}(t) = f(t, x(t))$; the expression is a Volterra integral equation representation of the solution. It is key to the proof of the existence and uniqueness theorem, but is not very useful since it gives no clue of what is $x(t)$.

Example. $\dot{x} = t^2 + x^2$, $x(0) = 1$, then $x(t)$ satisfies

$$x(t) = 1 + \int_0^t [s^2 + x^2(s)]ds,$$

which is not a terribly enlightening fact.

5 Existence and Uniqueness—Preliminaries

First of all, if the world's collection of ODEs were represented by this page, then this dot
• would be the ones that can be explicitly solved, meaning we can find an explicit
representation of a solution $x(t, C)$. This is why we need numerical and graphical
techniques, and mathematical analysis to approximate solutions or describe their properties
or behavior. But to do this we need to have some assurance that solutions exist.

The statement of existence depends on *initial values* (initial conditions), so we formulate
the *initial value problem*:

(IVP) $\dot{x} = f(t, x), \quad x(t_0) = x_0.$

Analogous to the implicit function theorem we want conditions on f assuring us that a
solution exists inside some region of the point (t_0, x_0).

We want to know that there exists at least one differentiable function $x(t)$, defined
near t_0, and satisfying $\dot{x}(t) = f(t, x(t))$, and passing through the point (t_0, x_0).

For simplicity suppose that B is a rectangle with center (t_0, x_0),

$$B = \{(t, x) \mid |t - t_0| \leq a, |x - x_0| \leq b\}.$$

Then, *existence* of a solution of the IVP is given by the theorem:

> Suppose that $f(t, x)$ is continuous in B. Then there exists a solution $x(t)$ of the
> IVP defined for $|t - t_0| \leq r \leq a$, with $r > 0$.

Pretty simple—but note the theorem is *local* in nature, since the solution is guaranteed
to exist only for $|t - t_0| \leq r$, and r might satisfy $r \ll a$. The rectangle B might be very
large yet r might turn out to be much smaller than its lateral dimension. This leads to the
continuation problem which we will briefly discuss after we discuss *uniqueness* which is
given by the statement:

> Suppose in addition that $\partial f(t, x)/\partial x$ is also continuous in B. Then the solution
> of the IVP is unique.

This means there is only one solution passing through (t_0, x_0), but just remember
uniqueness means "solutions don't cross!" so we can't have a picture like this:

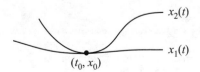

where $x_1(t), x_2(t)$ are both solutions of the IVP. From a physical, real world sense such a situation would be extremely unlikely, and would be a nightmare numerically, since most numerical schemes try to approximate the derivatives of solutions, and use them to venture forth from (t_0, x_0). But more importantly, the statement that solutions don't cross means that it can *never happen*. For if two solutions $x_1(t)$, $x_2(t)$ of $\dot{x} = f(t, x)$ intersected at any point (t_α, x_α), hence $x_1(t_\alpha) = x_2(t_\alpha) = x_\alpha$, then they would both be solutions of the IVP with initial conditions $x(t_\alpha) = x_\alpha$, and this can't occur. This last observation has important consequences in the study of the qualitative behavior of solutions.

Remark. A slightly more general condition that guarantees uniqueness is that $f(t, x)$ satisfy a Lipschitz condition in B:

$$|f(t, x) - f(t, y)| \leq K|x - y|, \qquad (t, x), (t, y) \text{ in } B$$

for some $K > 0$. But the condition of continuous partial derivative $\frac{\partial f}{\partial x}$ in B is much easier to verify. Example: Since $\big||x| - |y|\big| \leq |x - y|$ then $f(t, x) = |x|$ satisfies a Lipschitz condition with $K = 1$ in any B of the form $\{(t, x) \,\big|\, |t - t_0| \leq a, |x| \leq b\}$ but $\frac{\partial f}{\partial x}$ is discontinuous at $x = 0$. Nevertheless the IVP, $\dot{x} = |x|$, $x(t_0) = x_0$, has unique solutions for any initial value; if $x_0 = 0$ the solution is $x(t) \equiv 0$.

6   Continuation

This is not a subject easily discussed—the analysis is delicate—but the intuitive result carries one a long way. Go back to the existence and uniqueness theorem for the IVP: $\dot{x} = f(t, x)$, $x(t_0) = x_0$, and suppose we have found an r and a unique solution $x(t)$ defined for $|t - t_0| \leq r$.

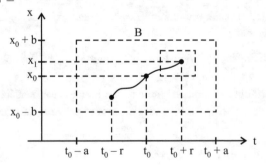

Since the point $(t_0 + r, x(t_0 + r)) = (t_0 + r, x_1)$ is in B, then $f(t, x)$ and $\partial f(t, x)/\partial x$ are continuous in some box B_1, possibly contained in B, or possibly extending beyond it, with center $(t_0 + r, x_1)$.

Now define a new IVP

$$\dot{x} = f(t, x), \quad x(t_0 + r) = x_1,$$

and the existence and uniqueness theorem guarantees a unique solution defined for $|t - (t_0 + r)| < r_1, r_1 > 0$, which will *agree* with $x(t)$ on their overlapping interval of existence. We have *continued* or extended the original solution $x(t)$ defined on the interval $t_0 - r \leq t \leq t_0 + r$ to the interval $t_0 - r \leq t \leq t_0 + r + r_1$.

It seems plausible that we can continue this game which will eventually stop and the solution will become unbounded or not be defined, or we can continue forever and the solution will be defined on an infinite time interval, $t < \alpha$, $t > \beta$, or $-\infty < t < \infty$. Furthermore, it seems equally plausible that for the IVP, the solution $x(t)$ satisfying $x(t_0) = x_0$ will possess a *maximum interval of existence*,

$$t_0 + \alpha < t < t_0 + \beta, \quad \alpha, \beta \text{ finite or infinite}$$

and that interval will depend on t_0 and x_0.

At this point there are several other topics and refinements regarding existence and uniqueness that should be discussed. I will defer them until the end of the next chapter— some readers may wish to go there directly, but some may want to wrap their hands around some honest to gosh solutions.

2

First Order Equations

The author has long been convinced that a study in depth of the first order, scalar, ODE is the key to understanding many of the qualitative properties of higher order equations, or systems of equations. But to spend a lot of time on arcane or rarely encountered equations (exact, Clairaut's, Bernoulli), or learning immediately forgotten transformations which convert some ugly ODE into a solvable, less ugly one will not likely develop that understanding. Borrow a copy of Kamke's encyclopedic book, *Differential Gleichgungen, Losungmethoden und Losungen* if you want to see a comprehensive analysis of that little dot mentioned previously.

The two key equations to be studied are

$$\frac{dx}{dt} = f(t)g(x), \qquad \text{The Separable Equation}$$
$$\frac{dx}{dt} = a(t)x + b(t) \quad \text{The Linear Equation}$$

and one should continuously remind oneself that a course in ordinary differential equations is *not* a post-sophomore course in techniques of integration. Following that philosophy or the one mentioned in the previous paragraph, or slavishly entering the ODE into the DSolve routine in the computer is what gives the subject the reputation of being a "bag of tricks."

1 The Separable Equation—Expanded

The ODE $\frac{dx}{dt} = f(t)g(x)$ can be broken down into three separate cases:

$$\text{A) } g(x) \equiv 1, \quad \text{B) } f(t) \equiv 1, \quad \text{C) neither } A \text{ nor } B.$$

For pedagogical reasons this makes good sense, since the level of difficulty goes up at each stage.

A. $\frac{dx}{dt} = f(t)$

Assume that $f(t)$ is continuous on some interval $a < t < b$, then the Existence and Uniqueness Theorem (hereafter referred to as E and U) gives us that B can be any rectangle contained in the strip $\{(t,x) \mid a < t < b, -\infty < x < \infty\}$ and a unique solution of the IVP

$$\frac{dx}{dt} = f(t), \quad x(t_0) = x_0, \qquad a < t < b,$$

exists. Let

$$x(t) = x_0 + F(t) = x_0 + \int_{t_0}^{t} f(s)ds$$

and verify that $x(t)$ is the desired solution, but note that we may not be able to explicitly evaluate the integral.

Examples.

a) $\frac{dx}{dt} = t^2 + \sin t, x(1) = 1$. Then $-\infty < t < \infty$ and $x(t) = 1 + \int_{1}^{t}(s^2 + \sin s)ds = \frac{t^3}{3} - \cos t + (\frac{2}{3} + \cos 1)$

b) $\frac{dx}{dt} = t\sqrt{1 - t^2}$, and E and U satisfied only for $|t| < 1$;

$$x(t) = x_0 + \frac{1}{3}(1 - t_0^2)^{3/2} - \frac{1}{3}(1 - t^2)^{3/2}$$

is the solution satisfying $x(t_0) = x_0$, $|t_0| < 1$.

c) $\frac{dx}{dt} = \sin t^2, x(0) = 4$. Then $-\infty < t < \infty$ and $x(t) = 4 + \int_{0}^{t} \sin^2 s \, ds$, but the integral can only be evaluated numerically (e.g., Simpson's Rule). To approximate, for instance, $x(1)$ you must evaluate

$$x(1) = 4 + \int_{0}^{1} \sin s^2 \, ds \approx 4.31027.$$

Going back to the expression for $x(t)$ and substituting t_1 for t_0 and x_1 for x_0 we see that the solution satisfying $x(t_1) = x_1$ can be given by

$$x(t) = \left(x_1 - \int_{t_0}^{t_1} f(s)ds\right) + \int_{t_0}^{t} f(s)ds = \left(x_1 - F(t_1)\right) + F(t),$$

so every solution is a translate of $F(t)$, or any other antiderivative, in the x-direction.

B. $\frac{dx}{dt} = g(x)$

Assume that $g(x)$ is continuously differentiable on some interval $c < x < d$, so $g(x)$ and $\frac{\partial g}{\partial x} = g'(x)$ are continuous there. Then B is any rectangle in the strip $\{(t,x) \mid -\infty < t < \infty, c < x < d\}$ and a unique solution of the IVP

$$\frac{dx}{dt} = g(x), \quad x(t_0) = x_0, \qquad c < x_0 < d,$$

exists.

An important case arises here—suppose that $g(x_0) = 0$. Then the constant solution $x(t) \equiv x_0$ is the unique solution of the IVP. Such a solution is an *equilibrium* or *singular solution*, and we should look for them all *before* embarking on the solution procedure. A further analysis of these solutions will be given in the next section.

Suppose $g(x_0) \neq 0$ then consider the expression

$$\int_{x_0}^{x} \frac{1}{g(r)} dr \equiv G(x) = t - t_0.$$

If $x = x_0$ then $G(x_0) = 0 = t - t_0$ so $t = t_0$. Furthermore, by the chain rule we obtain, after differentiating each side with respect to t:

$$\frac{d}{dt} G(x) = G'(x) \frac{dx}{dt} = \frac{1}{g(x)} \frac{dx}{dt} = \frac{d}{dt}(t - t_0) = 1.$$

Hence, the expression $G(x) = t - t_0$, gives an *implicit expression* for the solution $x(t)$ of the IVP.

In theory, since $G'(x) = 1/g(x) \neq 0$, then $G(x)$ is invertible and therefore

$$x(t) = G^{-1}(t - t_0).$$

The last expression is an elegant one, and since

$$\int_{x_0}^{x} \frac{1}{g(r)} dr - \int_{x_0}^{x_1} \frac{1}{g(r)} dr = \int_{x_1}^{x} \frac{1}{g(r)} dr = t - t_1$$

one sees that the solution for any initial value is given by $x(t) = G^{-1}(t + C)$ for an appropriate value of C. Every solution is a translate of $G^{-1}(t)$ in the t-direction.

There is a two-fold problem, however. It may not be possible to evaluate explicitly the integral $\int 1/g(r) \, dr$, and even if it is, one may not be able to invert $G(x)$ to get an explicit expression for $x(t)$. One can only resort to numerical or analytic (e.g., series) techniques to obtain approximations of the solution.

Important Remark. Having now satisfied the demands of rigorous analysis, one can now proceed more cavalierly in an effort to make computation easier.

Case A: $\frac{dx}{dt} = f(t)$. Write $dx = f(t)dt$ then integrate to get $x(t) = \int f(t)dt + C$ and choose C so that $x(t_0) = x_0$.

Case B: $\frac{dx}{dt} = g(x)$. Write $dx/g(x) = dt$ then integrate both sides to get $G(x) = t + C$ and choose C satisfying $G(x_0) = t_0 + C$. Invert if possible to get $x(t) = G^{-1}(t + G(x_0) - t_0)$; the last two steps can be interchanged if $G(x)$ can be inverted.

Examples.

a) $\frac{dx}{dt} = kx$, then $g(x) = kx$, $g'(x) = k$, and $-\infty < x < \infty$. Since $g(0) = 0$, $x(t) \equiv 0$ is an equilibrium so we must consider $x < 0$ or $x > 0$. Proceeding as above

$$\frac{dx}{kx} = dt \quad \text{or} \quad \frac{1}{k} \ln x = t + C,$$

then

$$x(t) = e^{kt+kC} = e^{kt}e^{kC} = Ce^{kt}.$$

Since C is an arbitrary constant we can take liberties with it, so e^{kC} becomes C, and $x(t) > 0$ implies $C > 0$, $x(t) < 0$ implies $C < 0$. We may have offended purists who insist that $\int \frac{1}{x}\,dx = \ln|x|$, but carrying around the absolute value sign is a nuisance, and the little C stratagem forgives all. Then $x(t_0) = x_0$ gives $C = x_0 e^{-kt_0}$ (so C takes the sign of x_0) and $x(t) = x_0 e^{k(t-t_0)}$ which is

exponential decay if $k < 0$ and $x(t) \to 0$, the equilibrium, as $t \to \infty$
exponential growth if $k > 0$ and $x(t)$ becomes unbounded as $t \to \infty$.

b) $\frac{dx}{dt} = 1 + x^2$, $g(x) = 1 + x^2$, $g'(x) = 2x$, and $-\infty < x < \infty$. Then

$$\int \frac{dx}{1+x^2} = \tan^{-1} x = t + C.$$

(Note we have played games with constants again, since to be calculus-proper we should write

$$\int \frac{dx}{1+x^2} = \tan^{-1} x + C = t + C_1.$$

But let $C_1 - C$ become C, thus easing bookkeeping). Then $x(t) = \tan(t + C)$ and if $x(t_0) = x_0$ we have $x(t) = \tan(t + (\tan^{-1} x_0 - t_0))$. The *real* importance of this equation is that the solution becomes infinite on a finite interval, $-\pi/2 - (\tan^{-1} x_0 - t_0) < t < \frac{\pi}{2} - (\tan^{-1} x_0 - t_0)$, despite the deceiving ubiquity and smoothness of $g(x) = 1 + x^2$. This is a very good example of *finite escape time*—the solution blows up at finite values of t. The interval of existence of a solution is limited in ways not apparent from the ODE itself.

c) $\frac{dx}{dt} = 3x^{2/3}$; this is the time honored representative of nonuniqueness of solutions of the IVP. Since $g(x) = 3x^{2/3}$ then $g(0) = 0$ so $x(t) \equiv 0$ is a solution, but since $g'(x) = 2x^{-1/3}$ is not continuous at $x = 0$, there may not be uniqueness, which is the case. If $x(t_0) = x_0$ then

$$\int_{x_0}^{x} \frac{dr}{3r^{2/3}} = x^{1/3} - x_0^{1/3} = \int_{t_0}^{t} dt = t - t_0$$

so $x(t) = (t - t_0 + x_0^{1/3})^3$, and if $x_0 = 0$ then $x(t) = (t - t_0)^3$. Both it and the solution $x(t) \equiv 0$ pass through the point $(t_0, 0)$. In fact, since $\frac{dx}{dt} = 3(t - t_0)^2$ equals 0 at $t = t_0$ we have *four* solutions passing through $(t_0, 0)$: $x(t) \equiv 0$, $x(t) = (t - t_0)^3$, and

$$x(t) = \left\{ \begin{array}{ll} 0, & t \le t_0 \\ (t - t_0)^3, & t > t_0 \end{array} \right., \qquad x(t) = \left\{ \begin{array}{ll} (t - t_0)^3, & t \le t_0 \\ 0, & t > t_0 \end{array} \right.$$

The agreement of the derivative at $(t_0, 0)$ of both solutions makes it possible to glue them together.

4) $\frac{dx}{dt} = \frac{1}{x^4+4}$ and $\frac{dx}{dt} = x^4 + 4$, $x(t_0) = x_0$. In both cases, $-\infty < x < \infty$, but in the first case we obtain the implicit representation

$$\frac{x^5}{5} + 4x = t - t_0 + \frac{x_0^5}{5} + 4x_0,$$

which can't be solved for $x = x(t)$, the solution. In the second case we get

$$\int_{x_0}^{x} \frac{1}{r^4 + 4} dr = t - t_0$$

and the integral cannot be explicitly evaluated. In both cases, you must call your friendly number cruncher.

C. $\frac{dx}{dt} = f(t)g(x)$

If $f(t), g(x)$ and $\frac{\partial}{\partial x}(f(t)g(x)) = f(t)g'(x)$ are continuous for $a < t < b$ and $c < x < d$, then unique solutions of the IVP are assured. First, find all values $x = k$ where g vanishes then $x(t) \equiv k$ are singular solutions. Otherwise, if

$$F(t) = \int f(t)dt, \quad G(x) = \int \frac{dx}{g(x)}$$

then an implicit expression for the solution $x(t)$, satisfying the IC, $x(t_0) = x_0$ is

$$G(x) = F(t) + G(x_0) - F(t_0).$$

Since $G'(x) = g(x) \neq 0$, then G is locally invertible and $x(t) = G^{-1}(F(t) + G(x_0) - F(t_0))$. All the joys and headaches of the previous case remain, and of course, we can "separate" variables to write the informal solution approach

$$\frac{dx}{g(x)} = f(t)dt, \text{ etc.}$$

Examples. Build your own, but here's a nice one showing how the maximum interval of existence of a solution can depend on the initial values:

$$\frac{dx}{dt} = -3x^{4/3}\sin t, \quad -\infty < t < \infty, -\infty < x < \infty.$$

Then $f(t) = \sin t$, $g(x) = -3x^{4/3}$ and $g'(x) = -4x^{1/3}$ are continuous, and

$$\frac{dx}{x^{4/3}} = -3\sin t \, dt \quad \text{implies}$$

$$-3x^{-1/3} = 3\cos t + C.$$

Letting C become $-3C$ gives the solution $x(t) = (C - \cos t)^{-3}$. If $|C| > 1$ then it is defined for $-\infty < t < \infty$, whereas if $|C| \leq 1$ it will only be defined on a finite interval. For instance

$$x(\pi/2) = \tfrac{1}{8} \text{ implies } C = 2, \quad x(t) = (2 - \cos t)^{-3}, \quad -\infty < t < \infty,$$

$$x(\pi/2) = 8 \text{ implies } C = 1/2, \quad x(t) = \left(\tfrac{1}{2} - \cos t\right)^{-3},$$

which is defined only for $\pi/3 < t < 5\pi/3$.

2 Equilibria—A First Look at Stability

We consider the ODE $\frac{dx}{dt} = g(x)$, where $g(x)$ and $g'(x)$ are continuous for x in some interval $a < x < b$, and suppose $g(x_0) = 0$, and x_0 is an isolated zero of g. Then

$x(t) \equiv x_0$ is a constant solution of the IVP

$$\frac{dx}{dt} = g(x), \quad x(t_0) = x_0,$$

for any t_0, and is defined for $-\infty < t < \infty$. We call such a solution an *equilibrium* or *singular solution*, but perhaps a better terminology is x_0 is a *critical point* of the ODE.

The key to the characterization of x_0 is the value of $g'(x_0)$.

$g'(x_0) > 0$:   Then $\frac{dx}{dt} = g(x)$ is negative for $x < x_0$ and near x_0, and is positive for $x > x_0$ and near x_0, since $g(x)$ is an increasing function near x_0 and vanishes at x_0. Consequently, any solution $x(t)$ with an initial value $x(t_0) = x_1$, $x_1 < x_0$ and x_1 near x_0, must be a decreasing function so it moves away from x_0. But if $x_1 > x_0$ and x_1 near x_0 the solution must be an increasing function and it moves away from x_0. We have the picture,

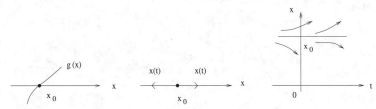

The middle picture is one of the *phase line* where we regard solutions $x(t)$ as moving points on the x-axis—it is a useful way to display the behavior of solutions in the one-dimensional case but the third picture is more illustrative. We can call such a critical point an *unstable equilibrium point or source* (a term inherited from physics). In physical terms we can think of the ODE as governing a system which is at rest when $x(t) = x_0$, but any small nudge will move the system away from rest. The solutions are repelled by their proximity to x_0 which gives rise to another colorful name: x_0 is a *repeller or repelling point*.

$g'(x_0) < 0$:   The previous analysis repeated leads to the following set of pictures:

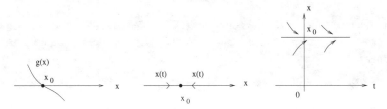

We call such a critical point a *stable equilibrium point or sink* (physics again lends a nice household name). In physical terms the system is at rest when $x(t) = x_0$, and under any small nudge the system will attempt to return to the rest position. Nearby solutions are attracted to x_0 which gives us another name: x_0 is an *attractor or attracting point*.

This is our first venture into the qualitative theory of ODEs, and the *important* point is that the analysis allows us to describe the behavior of solutions near x_0 whether we can explicitly solve the ODE or not!

Example. $\frac{dx}{dt} = \frac{x}{x-1}, x \neq 1$. Then $g(0) = 0$, and $g'(0) = -1$ so $x_0 = 0$ is a stable equilibrium point. Solving the equation leads to the implicit expression for $x = x(t)$: $x - \ln x = t + C$ which we can't solve for $x = x(t)$. But we have a very good picture of what happens to solutions whose initial values are close to 0.

The last example brings up a point worth amplifying. Suppose x_0 is a stable equilibrium point and $g'(x_0) < 0$. Then given any solution $x(t)$ with initial value $x(t_0) = x_1$, with $x_1 > x_0$ but sufficiently close to x_0, we have

$$\frac{dx}{dt}\bigg|_{t=t_0} = g(x_1) < 0 \quad \text{(go back to the relevant picture)}$$

hence $x(t)$ is decreasing near $t = t_0$. But now apply the existence and uniqueness criteria, and the continuation argument previously given, and we can infer that $x(t)$ must continue to decrease, but since it can't cross the line $x = x_0$, we conclude that $\lim_{t\to\infty} x(t) = x_0$. A rigorous argument justifying the last assertions, which are correct, might take a few paragraphs, but we should follow our geometric hunches—they are rarely wrong (frequently right?).

The last statement leads to the definition of the *asymptotic stability* of x_0, for which a homespun characterization (avoiding ϵs and δs) at this point is all that is needed.

> The equilibrium point x_0 is asymptotically stable if once a solution $x(t)$ is close to x_0 for some $t = t_1$, it stays close for all $t > t_1$, and furthermore $\lim_{t\to\infty} x(t) = x_0$.

To expand on this requires the subtle distinction between stability (once close, always close) and asymptotic stability. This distinction is a much richer one in the study of higher dimensional systems, and makes little sense for one-dimensional ODEs, unless we want to do a little nit-picking and discuss the case $g(x) \equiv 0$ when every x_0 is a solution, and so is every x_1 as close to x_0 as we want.

$g'(x_0) = 0$: This is the ambiguous case, and requires individual analysis of an ODE having the property. We can have

$$x_0 \text{ is stable:} \quad \text{e.g., } x_0 = 0 \quad \text{and} \quad g(x) = -2x^5,$$
$$x_0 \text{ is unstable:} \quad \text{e.g., } x_0 = 0 \quad \text{and} \quad g(x) = x^3;$$

the reader is left to verify the conclusions. But a third possibility arises in which x_0 is a *semistable equilibrium point*.

Example. $\frac{dx}{dt} = x^2$, so $g(0) = 0$, and $g'(0) = 0$. Solve the equation to obtain $x(t) = -(t + C)^{-1}$. Let $x(0) = -\epsilon$, $\epsilon > 0$ then $x(t) = -\left(t + \frac{1}{\epsilon}\right)^{-1}$ which approaches $x_0 = 0$ as $t \to \infty$. But let $x(0) = \epsilon$, $\epsilon > 0$ then $x(t) = -\left(t - \frac{1}{\epsilon}\right)^{-1}$ which is an increasing function for $t > 0$ and blows up when $t = \frac{1}{\epsilon}$.

A very annoying case, both mathematically and physically!

This is a very good time to introduce the notion of *linearized stability analysis* which is a formidable name for a simple application of Taylor's Formula, and will be more extensively used in the study of higher dimensional nonlinear systems.

If x_0 is an equilibrium or critical point of the ODE $dx/dt = g(x)$, then to study its stability we can consider a nearby solution $x(t)$ and write it as $x(t) = x_0 + \eta(t)$, where $\eta(t)$ is regarded as a *small* perturbation. Since $x_0 + \eta(t)$ is a solution then

$$\frac{d}{dt}(x_0 + \eta(t)) = 0 + \frac{d}{dt}\eta(t) = g(x_0 + \eta(t)),$$

and if $g(x)$ is sufficiently smooth we can apply Taylor's Formula, recalling that $g(x_0) = 0$. We get

$$\frac{d\eta}{dt} = 0 + g'(x_0)\eta + g''(x_0)\frac{\eta^2}{2!} + \text{(higher order terms)}$$

and the *key assumption* is that the quadratic and higher order terms, which are very small if $|\eta(t)|$ is small, have little or no effect, and the dominant equation governing the growth of $\eta(t)$ is the linear approximation

$$\frac{d\eta}{dt} = g'(x_0)\eta \quad \text{or} \quad \eta(t) = C\exp\left[g'(x_0)t\right].$$

We see immediately that

$g'(x_0) < 0$ implies $\eta(t) \to 0$ as $t \to \infty$ hence x_0 is stable,

$g'(x_0) > 0$ implies $\eta(t)$ becomes unbounded as $t \to \infty$ hence x_0 is unstable.

To analyze stability in the semistable case we have to find the first nonzero value $g^{(n)}(x_0)$ and solve the resulting nonlinear ODE. But for the case $g'(x_0) \neq 0$ the linearized analysis backs up our previous discussion, and will be crucial when we discuss equilibria of higher order systems, when dimensionality takes away our nice one-dimensional, calculus approach.

3   Multiple Equilibria

For a start consider the equation $\frac{dx}{dt} = g(x)$ where $g(x_1) = g(x_2) = 0$ and $x_1 < x_2$; for instance, $g(x) = x(x-1)$ and $x_1 = 0$, $x_2 = 1$. We will assume that $g(x)$ and $g'(x)$ are continuous so E and U of the solution of the IVP is assured. Thus, we have the two solutions $x_1(t) \equiv x_1$, $x_2(t) \equiv x_2$, and the picture

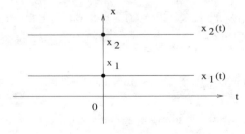

The most important conclusion we can draw from the picture is

Every solution $x(t)$ with initial conditions $x(t_0) = x_0$, where $x_1 < x_0 < x_2$, is defined for $-\infty < t < \infty$.

This is an important but frequently overlooked fact.[1]

The reasons for this conclusion are the continuation property and uniqueness of solutions. The existence and uniqueness result tells us that the solution will be defined for some interval $t_0 - r \le t \le t_0 + r$, $r > 0$, and for that interval the solution cannot intersect the solutions $x_1(t)$ or $x_2(t)$ by uniqueness. That means that the end points $x(t_0 - r)$ and $x(t_0 + r)$ will be inside the interval $x_1 < x < x_2$, and we can apply the existence and uniqueness theorem and further extend $x(t)$, over and over again. The solution $x(t)$ is trapped between the bounding solutions x_1 and x_2 and no matter how it wiggles it can never escape, but must move inexorably forward and backward. This is a wonderful result, since determining whether solutions exist for all time and are bounded is a tricky matter:

> $\frac{dx}{dt} = \cos x - x^2$, and $g(x) = 0$ at $x \approx \pm 0.82413$ so every solution with initial values $x(t_0) = C$, $|C| < 0.82413$ is defined for $-\infty < t < \infty$ and satisfies $|x(t)| < 0.82413$.

But we can go a little further and discuss stability, in the case of two equilibria, x_1 and x_2, $x_1 < x_2$.

Since $g(x)$ does not change sign between x_1 and x_2 it must be either

positive: in which case $\frac{dx}{dt} > 0$, so any solutions $x(t)$ between x_1 and x_2 must be *strictly* increasing, but remember "solutions don't cross." Consequently $\lim\limits_{t \to \infty} x(t) = x_2$.

or

negative: the same line of reasoning tell us that any solution between x_1 and x_2 must satisfy $\lim\limits_{t \to \infty} x(t) = x_1$.

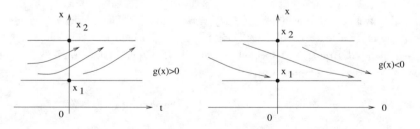

Note: that we only need to check the sign of $g(x)$ at any point between x_1 and x_2 which makes the job even easier.

> In the example above $g(0) = 1 > 0$ so we conclude any solution $x(t)$ with initial value $x(t_0) = C$, $|C| < 0.82413$ satisfies $\lim\limits_{t \to \infty} x(t) = 0.82413$.

Now what remains is to examine the behavior of solutions with initial values x_0 less than x_1 or greater than x_2. But since $g(x)$ cannot change sign for $x < x_1$ or $x > x_2$ we will get pictures like:

---

[1] At no cost you have discovered that the ODE has bounded solutions!

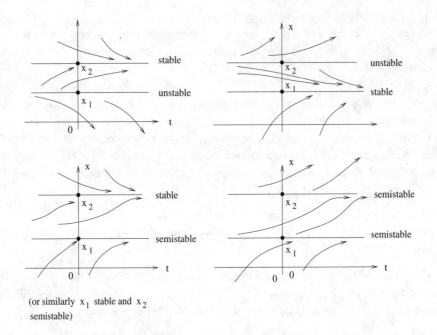

(or similarly x_1 stable and x_2 semistable)

The semistable case would be troublesome for any physical model; a solution could be close to an equilibrium value but a small change in the data could make it veer off.

A useful and popular model of the first picture above is the logistic equation of population growth:

$$\frac{dx}{dt} = rx\left(1 - \frac{x}{K}\right); \qquad \begin{aligned} r &= \text{ intrinsic growth rate } > 0, \\ K &= \text{ carrying capacity } > 0, \end{aligned}$$

with equilibria $x_1 = 0$ (unstable), and $x_2 = K$ (stable). The model is discussed in many of the current textbooks, as well as the variation which includes harvesting, first discussed by Fred Brauer and the author in a 1975 paper in the journal *Theoretical Population Biology*. That equation is

$$\frac{dx}{dt} = rx\left(1 - \frac{x}{K}\right) - H, \quad H = \text{ harvesting rate } > 0,$$

and a little calculus and some analysis shows that as H increases from 0, the unstable equilibrium increases from 0, and the stable equilibrium decreases from K. When H reaches $rK/4$ they both coalesce and for $H > rK/4$ we have $\frac{dx}{dt} < 0$ for all x so the population expires in finite time. For devotees of bifurcation theory the behavior can also be analyzed from the standpoint of a cessation of stability when the parameter H reaches the critical value $rK/4$: a saddle-node bifurcation.

One can now pile on more equilibria and look at equations like

$$\frac{dx}{dt} = (x - x_1)(x - x_2) \cdots (x - x_n)$$

and put them on the computer and look at direction fields, a pleasant, somewhat mindless exercise. Things can get wacky with equations like

$$\frac{dx}{dt} = \sin \frac{1}{x},$$ with an infinite number of equilibria $x_n = \pm \frac{1}{n\pi}$, $n = 1, 2, \ldots,$ converging to 0 and alternating in stability, or

$$\frac{d\theta}{dt} = \sin^2 \theta,$$ with an infinite number of semistable equilibria $x_n = 0, \pm\pi,$ $\pm 2\pi, \ldots,$ since in the vicinity of any equilibrium $\frac{d\theta}{dt} > 0.$

The last example will come up later in the discussion of the phase plane.

The case where the right-hand side of the differential equation depends on t and x (called the nonautonomous case) is much trickier, and while there are some general results, the analysis is usually on a case by case basis. The following example is illustrative of the complexities.

Example. $\dfrac{dx}{dt} = \dfrac{t(1 - x^2)}{x(1 + t^2)}, x \neq 0.$

We see that $x_1(t) \equiv 1$, $x_2(t) \equiv -1$ are solutions, so by uniqueness all other solutions lie between -1 and 1, are below -1 or are above 1. Now we must do some analysis:

Case: $0 < |x| < 1$. Then $\frac{dx}{dt} = 0$ when $t = 0$ and

$$\frac{dx}{dt} > 0 \quad \text{if} \begin{cases} t > 0 \\ x > 0 \end{cases} \qquad \frac{dx}{dt} < 0 \quad \text{if} \begin{cases} t < 0 \\ x > 0 \end{cases}$$

$$\frac{dx}{dt} > 0 \quad \text{if} \begin{cases} t < 0 \\ x < 0 \end{cases} \qquad \frac{dx}{dt} < 0 \quad \text{if} \begin{cases} t > 0 \\ x < 0 \end{cases}$$

Case: $|x| > 1$. Then $\frac{dx}{dt} = 0$ when $t = 0$ and

$$\frac{dx}{dt} > 0 \quad \text{if} \begin{cases} t > 0 \\ x > 1 \end{cases} \qquad \frac{dx}{dt} < 0 \quad \text{if} \begin{cases} t < 0 \\ x > 1 \end{cases}$$

$$\frac{dx}{dt} > 0 \quad \text{if} \begin{cases} t > 0 \\ x < -1 \end{cases} \qquad \frac{dx}{dt} < 0 \quad \text{if} \begin{cases} t < 0 \\ x < -1 \end{cases}$$

Since the equation is invariant under the change of variable $t \to -t$ we can conclude that

If $x(t_0) = x_0 > 0$, then $\lim_{t \to \pm\infty} x(t) = 1$, and

if $x(t_0) = x_0 < 0$, then $\lim_{t \to \pm\infty} x(t) = -1$.

Solutions have a maximum or minimum at $x(0)$ which becomes cusp-like as $x(0)$ gets closer to zero. (See diagram.) The effect of t on the asymptotic behavior of solutions is quite distinct from the autonomous (no t-dependence) case when solutions are strictly monotonic.

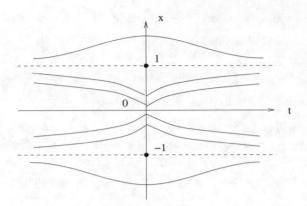

Admittedly, all but the last analysis could be done on a one-dimensional phase line, but one loses the attractive geometry of solutions moving around in the (t, x) plane. Furthermore, the approach taken is good training for later discussion of the phase plane, when solutions become trajectories carrying parametric representations of solutions.

4 The Linear Equation

We begin with the homogeneous linear equation

$$\dot{x} = a(t)x, a(t) \text{ continuous on } p < t < q,$$

which is a separable equation whose solution is

$$x(t) = x_0 \exp\left[\int_{t_0}^{t} a(s)ds\right]$$

satisfying the initial conditions $x(t_0) = x_0$. If we define $\Phi(t) = \exp\left[\int_{t_0}^{t} a(s)ds\right]$ then we can express the solution as $x(t) = x_0\Phi(t)$. This is the one-dimensional version of the solution of the n-dimensional linear system $\dot{x} = A(t)\underset{\sim}{x}$ where $\underset{\sim}{x} = \text{col}(x_1, \ldots, x_n)$ and $A(t) = (a_{ij}(t))$ is an $n \times n$ matrix with continuous components. The solution is $\underset{\sim}{x}(t) = \Phi(t)\underset{\sim}{x_0}$ where $\Phi(t)$ is a fundamental matrix satisfying the matrix ODE $\dot{\Phi} = A(t)\Phi$, $\Phi(t_0) = I$.

The fundamental matrix $\Phi(t)$ is a solution of the matrix ODE $\dot{\Phi} = A(t)\Phi$. In the case where $A(t) = A$ we have the series representation of the matrix exponential

$$\Phi(t) = \exp[tA] = I + tA + \frac{t^2}{2!}A^2 + \frac{t^3}{3!}A^3 + \cdots$$

which converges, and is a nice formal representation, but nobody in their right mind would compute it this way to solve the matrix ODE.

If we consider the nonhomogeneous linear equation

$$\dot{x} = a(t)x + b(t), a(t), b(t) \text{ continuous on } p < t < q,$$

then the solution satisfying $x(t_0) = x_0$ is given by

$$(*) \qquad x(t) = \exp\left[\int_{t_0}^{t} a(s)ds\right]\left[x_0 + \int_{t_0}^{t} \exp\left[-\int_{t_0}^{s} a(r)dr\right]b(s)ds\right].$$

The representation is easily obtained using the *very important variation of parameters* technique. Let $x(t) = \exp\left[\int_{t_0}^{t} a(s)ds\right]u(t) = \Phi(t)u(t)$, where we hope to find $u(t)$. If the initial conditions are $x(t_0) = x_0$ then $u(t_0) = x_0$ since $\Phi(t_0) = 1$. Substitute into the ODE to get

$$\dot{x} = \Phi\dot{u} + \dot{\Phi}u = a(t)\Phi u + b(t),$$

and since $\dot{\Phi} = a(t)\Phi$ we cancel to obtain

$$\Phi\dot{u} = b(t) \quad \text{or} \quad \dot{u} = \Phi^{-1}(t)b(t).$$

This is solved immediately

$$u(t) = x_0 + \int_{t_0}^{t} \Phi^{-1}(s)b(s)ds,$$

and since $\Phi^{-1}(s) = \exp\left[-\int_{t_0}^{s} a(r)dr\right]$ we get $(*)$.

But note the very important feature of the proof above, namely that $\Phi(t)$ could be instead the fundamental matrix of the n dimensional linear system $\dot{x} = A(t)x$, satisfying $\Phi(t_0) = I$, the identity matrix. The solution of the nonhomogeneous linear system $\dot{x} = A(t)x + B(t)$, where $B(t)$ is an n-dimensional continuous vector, is given by

$$(*)_n \qquad\qquad x(t) = \Phi(t)\left[x_0 + \int_{t_0}^{t} \Phi^{-1}(s)B(s)ds\right]$$

which is simply the n-dimensional version of $(*)$.

What is the point of escalating to this level of generality now? It is because almost all introductory texts write the first order scalar nonhomogeneous linear equation as

$$\dot{x} + a(t)x = b(t), \quad a(t), b(t) \text{ continuous on } p < t < q,$$

so they can use the traditional ploy of *integrating factors*. Note that the above expression is out of line with the universally accepted way to write the n-dimensional version as $\dot{x} = A(t)x + B(t)$.

The integrating factor approach is to multiply the equation by the integrating factor $\exp\left[\int_{t_0}^{t} a(s)ds\right]$, then

$$\exp\left[\int_{t_0}^{t} a(s)ds\right]\dot{x} + \exp\left[\int_{t_0}^{t} a(s)ds\right]a(t)x = \exp\left[\int_{t_0}^{t} a(s)ds\right]b(t).$$

One sagaciously notes that the left-hand side can be expressed as an exact derivative and we get

$$\frac{d}{dt}\left(\exp\left[\int_{t_0}^{t} a(s)ds\right]x\right) = \exp\left[\int_{t_0}^{t} a(s)ds\right]b(t).$$

Consequently,

$$\exp\left[\int_{t_0}^{t} a(s)ds\right] x(t) = x_0 + \int_{t_0}^{t} \exp\left[\int_{t_0}^{s} a(r)dr\right] b(s)ds,$$

and solving for $x(t)$ gives the expression $(*)$, but with $a(t)$ *replaced by* $-a(t)$.

This last remark is a source of confusion to many neophytes, since they first study the homogeneous equation $\dot{x} = a(t)x$, not $\dot{x} - a(t)x = 0$, then advance to $\dot{x} + a(t)x = b(t)$. Consequently, the sign change can cause problems in choosing the "right" integrating factor.

Secondly, the integrating factor approach has no analogue for higher dimensional systems. Write $\underset{\sim}{x} = A(t)\underset{\sim}{x} + B(t)$ as $\underset{\sim}{\dot{x}} - A(t)\underset{\sim}{x} = B(t)$, then you must multiply by $\Phi^{-1}(t)$ to obtain

$$\Phi^{-1}(t)\underset{\sim}{\dot{x}} - \Phi^{-1}(t)A(t)\underset{\sim}{x} = \Phi^{-1}(t)\underset{\sim}{B}(t).$$

Now you have to figure out whether the left-hand side equals $\frac{d}{dt}(\Phi^{-1}(t)\underset{\sim}{x})$ when all you know is that $\Phi(t)$ satisfies the matrix ODE $\dot{\Phi} = A(t)\Phi$. Tricky business!

You may not be convinced by the passionate argument above to eschew the "Find the integrating factor!" trick, and that's okay. Remembering $(*)$ means memorizing a formula, something held in low repute by some of today's mathematical educators. We will see that not using $(*)_n$ in developing the variation of parameters formula for higher order scalar ODE's makes life harder and introduces some algebraic complications. But choose your own petard.

Now let's turn to solutions. The formulas are there and if $\int^t a(t)dt$ has a known antiderivative then we can find an explicit solution of $\dot{x} = a(t)x$. Furthermore, if the same is true of $\int^t \exp\left[-\int^s a(r)dr\right] b(s)ds$ then the nonhomogeneous problem can be solved using $(*)$. Given the initial value $x(t_0) = x_0$ we can either incorporate it by using the definite integral $\int_{t_0}^{t}$, or just use the indefinite integral and an arbitrary constant to be evaluated later. The first approach is a little more efficient and reduces the chance of algebraic errors. If either or both of the definite integrals cannot be evaluated then we must resort to numerical methods.

The *problem* with many presentations of the solving of the nonhomogeneous equation is that the integral $\int_{t_0}^{t} \exp\left[\int_{t_0}^{s} a(r)dr\right] b(s)ds$ is used as a test of the student's ability to use (sometimes obscure) techniques of integration in what are obviously rigged problems. A relatively simple example is

$$\dot{x} = (\tan t)x + e^{\sin t}, \qquad 0 < t < \pi/2,$$

which the reader can work out, and note that if $\tan t$ were changed to $-\tan t$ the problem is intractable. It pays to keep in mind that the subject being studied is ordinary differential equations not integral calculus.

It is extremely important to emphasize that $(*)$ is a representation of the very important fact, true for all linear systems—the solution $x(t)$ is the sum of

$$\exp\left[\int_{t_0}^{t} a(s)ds\right] x_0 - \text{the general solution of the homogeneous}$$

equation $\dot{x} = a(t)x$, and

$$x_p(t) = \exp\left[\int_{t_0}^{t} a(s)ds\right] \int_{t_0}^{t} \exp\left[-\int_{t_0}^{s} a(r)dr\right] b(s)ds - \text{a particular}$$

solution of the nonhomogeneous equation.

The effort to find $x_p(t)$ must be an expeditious one, and almost all textbooks fail to point out that the *method of undetermined coefficients* (or judicious guessing), extolled for higher order constant coefficient linear equations, works for first order equations as well.

Consider the simple IVP

$$\dot{x} = 2x + 3t^2 - t + 4, \qquad x(0) = 4.$$

Using the expression for $x_p(t)$ above we obtain

$$x_p(t) = e^{2t} \int_0^t e^{-2s}(3s^2 - s + 4)ds$$

which will involve at least three integrations, mostly by parts, unless one happens to have a table of integrals or some sophisticated software handy.

But all the integrals will have a factor e^{-2t} which will be cancelled by the term e^{2t}, and one is left with a polynomial of degree 2. So why not try $x_p(t) = At^2 + Bt + C$, substitute into the ODE and solve for the arbitrary constants? Proceed:

$$\dot{x}_p = 2At + B = 2(At^2 + Bt + C) + 3t^2 - t + 4,$$

then equating like powers of t gives

$$0 = 2A + 3, \quad 2A = 2B - 1, \quad B = 2C - 4.$$

Solving these we obtain the general expression for $x(t)$

$$x(t) = x_0 e^{2t} - \frac{3}{2}t^2 - t + \frac{3}{2},$$

and $x(0) = 4$ gives $x_0 = 5/2$. Simplicity itself!

In the above problem, if the polynomial term were replaced by $a\sin qt$ or $b\sin qt$ the trial solution would be $x_p(t) = A\sin qt + B\cos qt$—you must include both. If instead the term were ae^{kt} the trial solution would be $x_p(t) = Ae^{kt}$ except in the case where $k = 2$, when it would be $x_p(t) = Ate^{2t}$, since Ae^{2t} is already a solution of the homogeneous equation. Students (and paper graders) will appreciate these obviations of tedious integrations.

The subject of existence of periodic solutions, when $a(t)$ and $b(t)$ are periodic will be covered in a later section when the linear equation and more general equations will be discussed. But an interesting topic is the case where $b(t)$, the forcing or nonhomogeneous term, is discontinuous. For instance, $b(t)$ could be a pulse which is turned on for a certain time interval, then turned off. Although the E and U theorem for $\dot{x} = f(t, x)$ requires that f be continuous in t, for the linear equation one sees that the integral expression

$$\int_{t_0}^t \exp\left[-\int_{t_0}^s a(r)dr \right] b(s)ds$$

is a *continuous* function of t, even if $b(t)$ is piecewise continuous. (One of my professors stated that integration make things nicer, whereas differentiation makes them worse—not a bad maxim). Consequently, the solution will be continuous, but have a discontinuous derivative.

The last observation takes us into the realms of higher analysis, where solutions are regarded as absolutely continuous functions, but this does not mean we need to ignore the question since it provides some nice problems.

Consider the general case with one discontinuity; more can be considered but this just involves extra bookkeeping. For simplicity, we let $a(t) = a$ since the expression $\exp\left[\int_{t_0}^t a(s)ds \right]$ plays no real role in the discussion. We can also let $t_0 = 0$, therefore

$$\dot{x} = ax + b(t), \quad x(0) = x_0 \quad \text{where}$$

$$b(t) = \begin{cases} b_1(t), & \text{continuous}, \ 0 \le t \le t_1 \\ b_2(t), & \text{continuous}, \ t_1 < t, \ \text{and } b_1(t_1) \ne b_2(t_1+), \end{cases}$$

so $b(t)$ has a jump discontinuity at $t = t_1$.

If we use $(*)$ and follow our noses we get the following expression for the solution:

$$x(t) = x_0 e^{at} + e^{at} \int_0^t e^{-as} b_1(s)ds, \quad 0 \le t \le t_1,$$

$$x(t) = x_0 e^{at} + e^{at} \int_0^{t_1} e^{-as} b_1(s)ds + e^{at} \int_{t_1}^t e^{-as} b_2(s)ds, \quad t_1 < t.$$

The integral expression is self-evident and what it really does is give us a method of solution which is:

Solve the IVP $\dot{x} = ax + b_1(t)$, $x(0) = x_0$, $0 \le t \le t_1$, and compute $x(t_1) = x_1$.

Solve the IVP $\dot{x} = ax + b_2(t)$, $x(t_1) = x_1$, $t_1 < t$, then glue the two solutions together at $t = t_1$.

We can use $(*)$ and the relation $e^{at} e^{-as} = e^{a(t-s)}$ to write a compact expression for the solution

$$x(t) = x_0 e^{at} + \int_{t_0}^t e^{a(t-s)} b(s)ds.$$

This is the elegant convolution integral representation of the solution, which is of greater interest when we discuss higher order equations; it is unfortunately usually never covered until one discusses Laplace transform theory.

Simple example: $\dot{x} = 2x + b(t), x(0) = 4, b(t) = \begin{cases} -1, & 0 \le t \le 1 \\ 1, & t > 1 \end{cases}$. Then

$$x(t) = 4e^{2t} + \int_0^t e^{2(t-s)}(-1)ds = \frac{7}{2}e^{2t} + \frac{1}{2}, \qquad 0 \le t < 1,$$

$$x(t) = 4e^{2t} + \int_0^1 e^{2(t-s)}(-1)ds + \int_1^t e^{2(t-s)}(1)ds$$

$$= \frac{7}{2}e^{2t} + e^{2(t-1)} - \frac{1}{2}, \quad t > 1.$$

We see that the two solutions agree at $t = 1$, but

$$\frac{dx}{dt}\bigg|_{t=1-} = 7e^2, \quad \frac{dx}{dt}\bigg|_{t=1+} = 7e^2 + 2.$$

The second expression for $x(t)$ can also be obtained by solving $\dot{x} = 2x + 1$, $x(1) = \frac{7}{2}e^2 + \frac{1}{2}$.

Finally, in keeping with this book's philosophy that the first order equation is an excellent platform to introduce more general topics at an elementary level, the author would like to discuss briefly the subject of *singular perturbations*. This will be done via a simple example which gives a glimpse into this fascinating topic of great interest in applied mathematics. For a more in-depth discussion see the references by Robert O'Malley.

Consider the IVP

$$\epsilon\dot{x} = -x + 1 + t, \quad x(0) = 0$$

where $\epsilon > 0$ is a small parameter.

The solution is

$$x(t) = (\epsilon - 1)e^{-t/\epsilon} + t + (1 - \epsilon)$$

and we see that it has a discontinuous limit as $\epsilon \to 0+$

$$\lim_{\epsilon \to 0+} x(t) = \begin{cases} 0, & t = 0 \\ 1 + t, & t > 0 \end{cases}.$$

Furthermore, if one is familiar with the "big O" notation, then

$$\dot{x}(t) = \left(\frac{1}{\epsilon} - 1\right)e^{-t/\epsilon} + 1 \approx \frac{1}{\epsilon} \quad \text{for} \quad 0 < t < O(\epsilon),$$

so what we have is a thin $O(\epsilon)$ region of transition, called a *boundary layer*, in which the solution rises rapidly from (0,0) and then approximates $x(t) = 1 + t$.

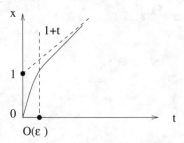

The function $x(t) = 1 + t$ obtained by setting $\epsilon = 0$ in the ODE is called the *outer solution*. The phenomenon described is very important in fluid mechanics where it models the flow of a liquid in a pipe near the walls of the pipe.

Another interesting example is the IVP

$$\epsilon \dot{x} = -(\sin t \cos t)x, \quad x(0) = 1,$$

and the points $t = n\pi/2$, $n = 0, \pm 1, \pm 2, \ldots$, where $a(t) = \sin t \cos t = \frac{1}{2}\sin 2t$ vanishes are called *turning points*. As before, $\epsilon > 0$ and small, and the solution is

$$x(t) = \exp(-(\sin^2 t)/2\epsilon).$$

The solution is negligible except at every other turning point $t = 0, \pm \pi, \pm 2\pi, \ldots$, where it has steep spikes of height 1 and width $O(\sqrt{\epsilon})$. For instance, $x(0) = 1$ and $\dot{x}(t) = -\frac{\sin 2t}{2\epsilon}\exp[-(\sin^2 t)/2\epsilon]$, and since $\sin\theta \approx \theta$ for $|\theta|$ small we see that if $0 < t < O(\sqrt{\epsilon})$ then $\dot{x}(t) \approx -O(1/\sqrt{\epsilon}) \rightarrow -\infty$ as $\epsilon \rightarrow 0+$.

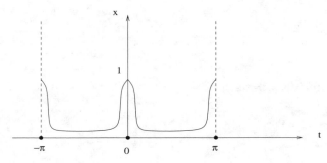

Unless one uses special numerical routines the solutions near the spikes will be tricky to compute.

5 The Riccati Equation

Although the reader was admonished to stay away from exotic special ODEs, the Riccati differential equation (first proposed by Count Jacopo Riccati, 1676–1754)[2] appears more

---

[2] The spelling of the first name was found on a picture the author has of the good Count. Sometime after 1754 the Italian language dropped entirely the *J* and replaced it with a Gi.

frequently in theoretical and applied analysis than the rest of the lot. Besides, it is the author's favorite equation.

The Riccati equation is a first order equation with a quadratic nonlinear term (the worst kind, some say), and is

$$\dot{x}(t) = p(t)x^2 + q(t)x + r(t),$$

where p, q, and are continuous on some interval $a < t < b$. It occurs in applications, some of which are:

a) The WKB method in asymptotic analysis: for the second order linear equation $\ddot{y} - \lambda^2 p(t)y = 0$, the transformation $x(t) = \dot{y}(t)/y(t)$ converts it to the Riccati equation $\dot{x} + x^2 = \lambda^2 p(t)$, and $y(t) = C \exp\left[\int^t x(s)ds\right]$.

b) The Riccati equation determines the feedback-control law for the linear regulator problem with quadratic cost. In the one-dimensional case this is the following: a system with output $x(t)$ and input or control $u(t)$ is governed by the linear ODE

$$\dot{x} = a(t)x + b(t)u, \quad x(t_0) = x_0, \quad t_0 \leq t \leq t_1.$$

The object is to choose a control $u = u(t)$ so as to minimize the cost functional or performance measure

$$C[u] = \frac{1}{2}kx(t_1)^2 + \frac{1}{2}\int_{t_0}^{t_1}\left[w_1(t)x(t)^2 + w_2(t)u(t)^2\right]dt,$$

where $w_1(t)$ and $w_2(t)$ are given weighting functions, and $k \geq 0$.

Optimal control theory gives that the solution is realized by the feedback control law:

$$u(t) = p^*(t)x(t), \quad p^*(t) = \frac{b(t)}{w_2(t)}p(t),$$

where $p(t)$ satisfies the Riccati equation

$$\dot{p} = -2a(t)p + \frac{b(t)^2}{w_2(t)}p^2 - w_1(t), \quad p(t_1) = k.$$

The control $u(t)$ is a feedback control since its present value is determined by the present value of the solution $x(t)$, via the feedback control law. A simple example of feedback control is the household thermostat. (A sample problem is given at the end of the section). For higher dimensional systems the Riccati equation is replaced by a matrix form of the equation.

c) The logistic equation, with or without harvesting, and with time dependent coefficients

$$\dot{x}(t) = r(t)x\left(1 - \frac{x}{K(t)}\right) - H(t),$$

is a Riccati equation. For instance, one could consider the case r, K constant and $H(t)$ periodic, corresponding to seasonal harvesting.

The last example will be discussed in a later section when we consider the general question of periodic solutions.

Riccati equations are notorious for having finite escape times, e.g., our old friend $\dot{x} = x^2 + 1$ with solutions $x(t) = \tan(t - C)$, and there is no general solution technique. But there are a few rays of hope. For instance, in the simpler case $r(t) \equiv 0$, the equation

$$\dot{x} = p(t)x^2 + q(t)x$$

is an example of a Bernoulli equation. The trick is to let $x(t) = 1/u(t)$ hence $\dot{x} = -\dot{u}/u^2$ and we obtain the linear equation

$$\dot{u} = -q(t)u - p(t),$$

which you may not be able to solve exactly, but at least have a formal expression for the solution. In the general case where $r(t) \not\equiv 0$ the transformation just gives you back another Riccati equation. Instead try the transformation

$$x(t) = -\frac{1}{p(t)}\frac{\dot{y}(t)}{y(t)},$$

and you obtain the second order linear equation

$$\ddot{y} - \left(q(t) + \frac{\dot{p}(t)}{p(t)} \right) \dot{y} + r(t)p(t)y = 0,$$

which probably won't make life much easier.

But suppose by a stroke of luck, or divine intervention, you can find or are given *one solution* $x(t) = \phi(t)$. Then let $x(t) = \phi(t) + 1/u(t)$, and substitute:

$$\dot{\phi} - \frac{\dot{u}}{u^2} = p(t)\left[\phi^2 + \frac{2\phi}{u} + \frac{1}{u^2} \right] + q(t)\left(\phi + \frac{1}{u} \right) + r(t)$$

which simplifies to the linear equation

$$\dot{u} = -\bigl(-2p(t)\phi(t) + q(t)\bigr)u - p(t),$$

since $\phi(t)$ satisfies the original equation. Here are several examples; the second one sheds further light on the notion of stability:

a) $\dot{x} = x^2 + 1 - t^2$ and a little staring reveals that $\phi(t) = t$ a solution. Letting $x(t) = t + \frac{1}{u}$ gives

$$1 - \frac{\dot{u}}{u^2} = \left(t + \frac{1}{u} \right)^2 + 1 - t^2$$

which simplifies to $\dot{u} = -2tu - 1$ with general solution

$$u(t) = e^{-t^2}\left[C - \int^t e^{r^2}\,dr \right].$$

If we specify initial conditions $x(0) = x_0$, then $x(t) = \phi(t) = t$ is the solution if $x_0 = 0$. Otherwise

$$x(t) = t + \frac{1}{u(t)} = t + \frac{e^{t^2}}{\frac{1}{x_0} - \int_0^t e^{r^2}\,dr}$$

and we observe distinct behavior depending on the value x_0, since the integral is a positive increasing function of t. If $x_0 < 0$, then solutions are defined for all t, whereas if $x_0 > 0$ then for some value $t = q > 0$ the denominator will equal zero, and the solution becomes infinite as t increases towards q.

Applying L'Hôpital's rule to the quotient we see that it approaches $-2t$ as $t \to \infty$ so $x(t)$ approaches $-t$; all solutions with $x(0) = x_0 < 0$ become asymptotic to the line $x(t) = -t$, which is not a solution:

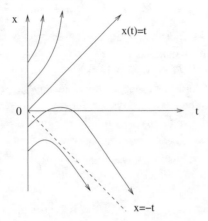

b) $\dot{x} = -x^2 + 2xt - t^2 + 5 = -(x-t)^2 + 5$. We see that the straight lines $x(t) = t + 2$, and $x(t) = t - 2$ are both solutions. By uniqueness, no solution can cross them, so we can conclude, for instance, that any solution $x(t)$ satisfying $|x(0)| \le 2$ is defined for all time. Letting $x(t) = t - 2 + \frac{1}{u}$ gives $\dot{u} = -4u + 1$, so

$$x(t) = t - 2 + \frac{1}{Ce^{-4t} + \frac{1}{4}},$$

which approaches $t + 2$ as $t \to \infty$. If $x(t) = t + 2 + \frac{1}{u}$ we get $\dot{u} = 2u - 1$ so

$$x(t) = t + 2 + \frac{1}{Ce^{2t} + \frac{1}{2}},$$

and $x(t)$ approaches $t + 2$ as $t \to \infty$. We can conclude that $x(t) = t + 2$ is asymptotically stable

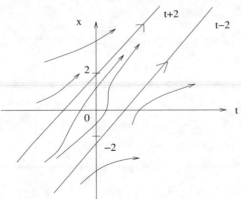

The above solution technique provides an amusing story. In the 1980s a firm entitled MACSYMA was widely advertising a symbolic and numeric computation package. One of their advertisements appeared frequently in the *American Mathematical Monthly* and showed four people in deep thought, staring at a blackboard—a clock on the wall hinted it was nearly quitting time. The title of the advertisement was "You can solve problems . . ." and on the blackboard was the initial value problem

$$\frac{dy}{dt} + y^2 + (2t + 1)y + t^2 + t + 1 = 0, \quad y(1) = 1.$$

The author noticed it was a Riccati equation, and after brief contemplation, that $y(t) = -t$ is a solution. Using the technique above led to a short calculation and the answer $y(t) = -t + (Ce^t - 1)^{-1}$, $C = 3/2e$, which blows up at $t = \ln(2e/3) \approx 0.5945$. No reply was ever received to a letter sent to the company suggesting they should provide more taxing problems to their prospective customers.

The subject of periodic solutions will be discussed in a later section, but this seems an excellent time to show a beautiful proof of a result for the Riccati equation, since it involves a special property of the equation, similar to the cross-ratio property of linear fractional transformations in conformal mapping theory. The proof can be found in the book by V. Pliss.

We are given solutions $x_i(t)$, $i = 1, 2, 3$ of a Riccati equation, hence they satisfy

$$\dot{x}_i = p(t)x_i^2 + q(t)x_i + r(t), \qquad i = 1, 2, 3.$$

A simple computation gives

$$\frac{\dot{x}_3(t) - \dot{x}_2(t)}{x_3(t) - x_2(t)} - \frac{\dot{x}_3(t) - \dot{x}_1(t)}{x_3(t) - x_1(t)} = p(t)(x_2(t) - x_1(t)),$$

and now suppose that

(i) $p(t), q(t), r(t)$ are periodic with minimum period T, and $\int_0^T p(s)ds \neq 0$; for simplicity let $p(t) \equiv 1$.

(ii) $x_i(t), i = 1, 2, 3$ are distinct T-periodic solutions, and we can assume $x_1(t) < x_2(t) < x_2(t)$ for all t.

Integrate the cross ratio equation from 0 to T and you obtain 0 on the left side and a nonzero quantity on the left. We conclude that

> The Riccati equation with T-periodic coefficients and $\int_0^T p(s)ds \neq 0$ can have at most two T-periodic solutions.

The condition on $p(t)$ is necessary; consider $\dot{x} = (\sin t)x^2$ which has an infinite family of solutions $x(t) = (c - \cos t)^{-1}$.

Remark. We will adopt the terminology that a function $x(t)$ is T-periodic, $T > 0$, if $x(t + T) = x(t)$ for all t, and T is the smallest such number.

The beautiful result above is a specific case of a more general question—given a polynomial differential equation of degree n with T-periodic coefficients,

$$\dot{x}(t) = x^n + a_{n-1}(t)x^{n-1} + \cdots + a_1(t)x + a_0(t),$$

what is the maximum number of T-periodic solutions it can have? For $n = 2$ we showed the answer is two; it has been shown that for $n = 3$ the answer is three. But there exist equations with $n = 4$ which have more than four T-periodic solutions, and considerable research has been done on the general problem.

Sample control theory problem: Given the linear regulator problem

$$\dot{x} = x + u, \quad x(0) = 1$$

and cost functional

$$C[u] = \frac{1}{2}kx(1)^2 + \frac{1}{2}\int_0^1 \left[3x(t)^2 + u(t)^2\right]dt:$$

a) Find the feedback control law, the optimal output $x(t)$, and the value of $C[u]$ for the case $k = 3$.

b) Find the feedback control law for the case $k = 0$, and use numerical approximations to find $u(x), x(t)$, and $C[u]$.

6 Comparison Sometimes Helps

This short section is merely to point out a very useful tool in the ODE mechanics toolbox—the use of comparison results to obtain some immediate insight into the behavior of a solution. These are often overlooked in introductory courses.

The simplest and most often used comparison result is the following one:

Given the functions $f(t, x)$, $g(t, x)$, and $h(t, x)$, all satisfying the existence and uniqueness criteria for the initial value problem in some neighborhood B of the point (t_0, x_0). Suppose that

$$f(t, x) < g(t, x) < h(t, x)$$

for (t, x) in B. Then the solutions of the initial value problem

$$\dot{y} = f(t, y), y(t_0) = x_0; \dot{x} = g(t, x), x(t_0) = x_0;$$
$$\dot{z} = h(t, z), z(t_0) = x_0,$$

satisfy the inequality

$$y(t) < x(t) < z(t)$$

for $t > t_0$ and (t, x) in B.

A formal proof can be given, but the intuitive proof observing that all the solutions start out at (t_0, x_0) and $\frac{dy}{dt} < \frac{dx}{dt} < \frac{dz}{dt}$ suffices. One or both sides of the inequality can be used.

Examples.

a) For $t \geq 1, x^2 + 1 < x^2 + t^2$ from which we can conclude that the solutions of $\dot{x} = x^2 + t^2$ have finite escape time. A sharper estimate would be that $x^2 < x^2 + t^2$

for $t > 0$ so the solutions of $\frac{dx}{dt} = x^2 + t^2$, $x(0) = x_0 > 0$ grow faster than the solutions of $\frac{dx}{dt} = x^2$, $x(0) = x_0$ which become infinite at $t = 1/x_0$.

b) Since

$$1 + t^2 < x^{1/2} + t^2 < x + t^2$$

for $t > 0$ and $x \geq 1$ we conclude that the solution of $\dot{x} = x^{1/2} + t^2$, $x(0) = 1$, is bounded for all time and satisfies the inequality

$$1 + t + t^2/3 < x(t) < 3e^t - t^2 - 2t - 2.$$

c) Some students presented the author with a problem in ocean climate modeling governed by the initial value problem

$$\dot{x} = \phi(t) - \epsilon x^4, \quad x(0) = 0, \quad \epsilon > 0,$$

where $\phi(t)$ was a periodic function too ugly to describe, but satisfied the inequality $0 \leq \phi(t) \leq k^2$. This implies that

$$-\epsilon x^4 \leq \phi(t) - \epsilon x^4 \leq k^2 - \epsilon x^4,$$

and since the solution of $\dot{x} = -\epsilon x^4$, $x(0) = 0$ is $x(t) \equiv 0$, and the equation $\dot{x} = k^2 - \epsilon x^4$ has two equilibrium solutions $x(t) = \pm\sqrt[4]{k^2/\epsilon}$, with the solution satisfying $x(0) = 0$ trapped in between, we can conclude that the solution of the IVP is bounded for all $t > 0$.

As the last example indicates, some comparison results are very handy, and a little analysis can pay off before embarking on some numerical work.

7   Periodic Solutions

A very deep question which occupies a sizeable part of the ODE literature is

> Given $\dot{x} = f(t, x)$, where f and $\frac{\partial f}{\partial x}$ are continuous for $-\infty < t < \infty$, $a < x < b$, and f is T-periodic, does the equation have a T-periodic solution?

By T-periodic, recall that we mean $f(t + T, x) = f(t, x)$ for all t, and T is the smallest such number, e.g., $\sin 2t$ has period $T = 2\pi/2 = \pi$ but obviously $\sin(2t + 2\pi) = \sin 2t$. Hence we are asking whether there exists one (or possibly many) solutions $x(t)$ satisfying

$$\dot{x}(t) = f(t, x(t)), \quad x(t + T) = x(t).$$

The extension of the question to higher dimensional systems $x \in R^n$, $f = (f_1, \ldots, f_n)$ is an obvious one.

For one-dimensional, autonomous systems $\dot{x} = f(x)$ it is obvious that they can't have any periodic solutions. An intuitive argument is that solutions move along a phase line unidirectionally so they can't return to a point they have passed. For higher dimensional systems this is not true: $\ddot{x} + x = 0$ is equivalent to the first order system $\dot{x} = y$, $\dot{y} = -x$, and every nontrivial solution is 2π-periodic. Therefore, we must consider one-dimensional, nonautonomous equations $\dot{x} = f(t, x)$.

Comment: with the current emphasis on dynamical systems, and the consequent geometry of trajectories and flows, plus the availability of excellent 2D or 3D graphics, there occasionally seems to be an unwillingness to discuss nonautonomous systems. They are sometimes avoided by increasing the dimension; for instance, the one-dimensional equation $\dot{x} = f(t, x)$ becomes the two-dimensional autonomous system $\dot{x} = f(y, x)$, $\dot{y} = 1$. Not a lot is gained except some nice graphs, and the author believes a study of the one-dimensional equation per sé gives a lot of insight into what happens at higher dimensions.

We start the discussion with the homogeneous linear equation

$$\dot{x} = a(t)x; \quad a(t) \text{ continuous}, \ -\infty < t < \infty; a(t + T) = a(t). \tag{L}$$

Since $f(t, x) = a(t)x$ and $\frac{\partial f}{\partial x}$ are continuous for all t and x, there is no loss of generality in assuming the initial conditions $x(0) = x_0$. Then the solution of (L) is

$$x(t) = x_0 \exp\left[\int_0^t a(s)ds\right], \text{ and}$$

by a change of limits of integration we get

$$x(t + T) = x_0 \left[\int_0^{t+T} a(s)ds\right] = x_0 \exp\left[\int_0^T a(s)ds\right] \exp\left[\int_T^{t+T} a(s)ds\right]$$

$$= x_0 \exp\left[\int_0^T a(s)ds\right] \exp\left[\int_0^t a(s)ds\right]$$

since $a(t + T) = a(t)$. But the last expression will equal $x(t)$ if and only if $\exp\left[\int_0^T a(s)ds\right] = 1$, and we conclude that

all solutions of (L) are T-periodic if and only if $\int_0^T a(s)ds = 0$.

This is the same as saying $a(t)$ has average value zero.

Example. $\dot{x} = (\sin t)x$ will have all 2π-periodic solutions whereas $\dot{x} = (1 + \cos t)x$ or $\dot{x} = |\sin t/2|x$ will have no 2π-periodic solutions.

Now what about the nonhomogeneous equation?

$$\dot{x} = a(t)x + b(t), \quad a(t), b(t) \text{ continuous}, \ -\infty < t < \infty,$$

$$a(t + T) = a(t), \quad b(t + T) = b(t). \tag{L$_n$}$$

The following is true:

(L)$_n$ will have a T-periodic solution for any $b(t)$ if and only if (L) has no nontrivial T-periodic solutions, or equivalently $\int_0^T a(s)ds \neq 0$.

The key to the proof is the observation that if a solution exists satisfying the relation $x(0) = x(T)$ then it must be periodic. For if $x(t)$ is a solution, then $y(t) = x(t + T)$ is also a solution since $a(t)$ and $b(t)$ are T-periodic. But $x(0) = x(T)$ implies $x(0) = y(0)$ so by uniqueness $x(t) = y(t) = x(t + T)$ for all t.

The proof of the assertion is to observe that the solution satisfying $x(0) = c$ is

$$x(t) = c \exp\left[\int_0^t a(s)ds\right] + \exp\left[\int_0^t a(s)ds\right]\int_0^t \exp\left[-\int_0^s a(r)dr\right]b(s)\,ds.$$

Then $x(0) - x(T) = 0$ implies that

$$c = \frac{\exp\left[\int_0^T a(s)ds\right]\int_0^T \exp\left[-\int_0^s a(r)dr\right]b(s)ds}{1 - \exp\left[\int_0^T a(s)ds\right]}$$

which has a unique solution since the denominator is not equal to zero.

Example. $\dot{x} = (1 + \cos t)x + \sin t$ and since $\int_0^{2\pi}(1 + \cos t)\,dt = 2\pi$ a periodic solution exists and $x(0) = x(2\pi) = c$ gives

$$c = \frac{\exp(2\pi)\int_0^{2\pi}\exp(-s - \sin s)\sin s\,ds}{1 - \exp(2\pi)} \approx -0.21088.$$

Numerically solving the IVP

$$\dot{x} = (1 + \cos t)x + \sin t, x(0) \approx -0.21088$$

gives $x(2\pi) \approx -0.20947$ which is good enough for a passing grade.

If we change $a(t)$ to $\sin t$ in the last example, no 2π-periodic solution will exist. The result for $(L)_n$ has generalizations to higher dimensional linear periodic systems.

Turning to nonlinear equations there are no general results, but a useful one will be given shortly to analyze a problem like the logistic equation with periodic harvesting

$$\dot{x} = rx\left(1 - \frac{x}{K}\right) - H(t), r > 0, K > 0 \quad \text{and} \quad H(t + T) = H(t).$$

We can assume $H(t) = H_0 + \epsilon \sin wt$, representing a small periodic harvesting rate, or $H(t) = \epsilon\phi(t)$ where $\phi(t + T) = \phi(t)$.

Recall that if $H(t) \equiv 0$, there are two equilibria $x(t) \equiv 0$ (unstable) and $x(t) \equiv K$ (stable), and if $H(t) \equiv H$ constant then $H > rK/4$ implies the population expires in finite time. So we might expect that if the periodic $H(t)$ is small then the equilibria might transmute into periodic solutions. Since the equation is a Riccati equation we know it can have at most two T-periodic solutions, which reinforces our intuition.

The result we need, found in some 1966 notes by K. Friedrichs, is the following:

Given $f(t, x)$, satisfying the conditions for E and U and $f(t+T, x) = f(t, x)$ for all (t, x), suppose there exist constants a, b, with $a < b$ such that $f(t, a) > 0$, $f(t, b) < 0$ for all t. Then there exists a T-periodic solution $x(t)$ satisfying $x(0) = c$, $a < c < b$.

A picture almost tells the whole story:

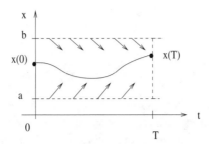

The solution viewed as a map takes the interval $\{a \le x \le b, t = 0\}$ continuously into the interval $\{a \le x \le b, t = T\}$, so by the Brouwer fixed point theorem there is a value c satisfying $x(0) = x(T) = c$. The corresponding solution extended periodically is a T-periodic solution.

Now rewrite the logistic equation as

$$\dot{x} = rx\left(1 - \frac{x}{K}\right) - H(t) = \frac{r}{K}\left[-\left(x - \frac{K}{2}\right)^2 + \left(\frac{K^2}{4} - \frac{KH(t)}{r}\right)\right]$$

and assume $0 < H(t) < rK/4$. Then

$$x = \frac{K}{2}: \quad \dot{x} = \frac{r}{K}\left[\frac{K^2}{4} - \frac{KH(t)}{r}\right] = \frac{rK}{4} - H(t) > 0$$

$$x = K: \quad \dot{x} = \frac{r}{K}\left[-\frac{K^2}{4} + \frac{K^2}{4} - \frac{KH(t)}{r}\right] = -H(t) < 0$$

and we can conclude there is a (stable) T-periodic solution $x(t)$ with $K/2 < x(t) < K$. Note also that

$$x = 0: \quad \dot{x} = -H(t) < 0$$

so if we change variables t to $-t$ we can conclude there is an (unstable) T-periodic solution $x(t)$ with $0 < x(t) < K/2$.

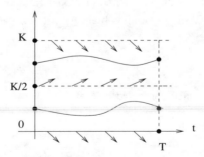

The reader might wish to generalize the result to the case where $r(t)$ and $K(t)$ are T-periodic and satisfy bounds like

$$0 < r_{\min} \le r(t) \le r_{\max} < \infty, \quad 0 < K_{\min} \le K(t) \le K_{\max} < \infty,$$

and $H(t) < r(t)K(t)/4$, replacing the constant a with $K(t)/2$ and b with $K(t)$. Or consider the case of proportional harvesting $H(t) = E(t)x$, $E(t+T) = E(t)$ and $0 < E_{min} \le E(t) \le E_{max} < r_{min}$.

Finally, a more general result which might be handy is the following one, proved by A.C. Lazer and the author in an article in *Mathematics Magazine*:

> Given $\dot{x} = g(x) + h(t)$, where $g(x)$ is a smooth function, and $h(t)$ is T-periodic, then if $g''(x)$ is either strictly positive or strictly negative for all x, there will exist at most two T-periodic solutions.

This verifies the previous result for the Riccati equation but allows for more generality.

Examples.

a) The logistic equation with periodic harvesting has $g(x) = rx(1 - x/K)$ and $g''(x) = -2r/K < 0$.

b) A more exotic example is $\dot{x} = 2e^{-x} - 2x + \sin wt$ and $g''(x) = 2e^{-x} > 0$. But $f(t, x)$ satisfies $1 \le f(t, 0) \le 3$, and $-2.264 \le f(t, 1) \le -0.264$ so there is one periodic solution $x(t)$ with $0 < x(0) < 1$. An examination of the direction field shows there can be no other.

8 Differentiation in Search of a Solution

If we are given the IVP

$$\frac{dx}{dt} = f(t, x) \quad x(t_0) = x_0$$

and a solution $x(t)$, then we immediately know that

$$\left. \frac{dx(t)}{dt} \right|_{t=t_0} = f(t_0, x_0)$$

and consequently a linear approximation to $x(t)$ would be $x(t) \approx x_0 + f(t_0, x_0)(t - t_0)$. This is the basis of Euler's Method of numerical approximation.

But if our tolerance for differentiation is high, we can use the chain rule to go further. Since

$$\frac{d^2 x}{dt^2} = \frac{d}{dt} f(t, x) = \frac{\partial f}{\partial x}\frac{dx}{dt} + \frac{\partial f}{\partial t} = \frac{\partial f}{\partial x}(t, x)f(t, x) + \frac{\partial f}{\partial t}(t, x),$$

then all these can be formally computed and evaluated at (t_0, x_0). This gives us an exact value of the second derivative of $x(t)$ at $t = t_0$ and allows us to get a quadratic approximation

$$x(t) \approx x_0 + f(t_0, x_0)(t - t_0) + \left. \frac{d^2 x}{dt^2} \right|_{t=t_0} (t - t_0)^2/2!$$

where

$$\left. \frac{d^2 x}{dt^2} \right|_{\substack{t=t_0 \\ x=x_0}} = \frac{\partial f(t_0, x_0)}{\partial x} f(t_0, x_0) + \frac{\partial f(t_0, x_0)}{\partial t}.$$

One could proceed to get higher order approximations, but the differentiations will get onerous in most cases.

The technique is the basis for the Taylor series numerical methods; for $n = 2$ it might be useful if you are stranded on a desert isle with only a legal pad, pen, and a ten dollar K-Mart calculator.

Example. $\frac{dx}{dt} = \frac{t^2}{1+\sqrt{x}}, x(1) = 4$, then

$$\left.\frac{dx}{dt}\right|_{\substack{t=1 \\ x=4}} = f(1,4) = \frac{1^2}{1+\sqrt{4}} = \frac{1}{3}$$

and since

$$\frac{\partial f}{\partial x} = \frac{-t^2}{2\sqrt{x}(1+\sqrt{x})^2}, \quad \frac{\partial f}{\partial t} = \frac{2t}{1+\sqrt{x}},$$

we get

$$\left.\frac{d^2x}{dt^2}\right|_{\substack{t=1 \\ x=4}} = \left(\frac{-1^2}{2\sqrt{4}(1+\sqrt{4})^2}\right)\left(\frac{1^2}{1+\sqrt{4}}\right) + \frac{2(1)}{1+\sqrt{4}} = \frac{71}{108}.$$

So

$$x(t) \approx 4 + \frac{1}{3}(t-1) + \frac{71}{108}\frac{(t-1)^2}{2!}.$$

An approximate value of $x(2)$ is 4.66204 and a highly accurate numerical scheme gives 4.75474.

Computing higher derivatives is more effective when used formally, and an example is the logistic equation where $f(x) = rx(1 - x/K)$. Then

$$\frac{d^2x}{dt^2} = \frac{df}{dx}\frac{dx}{dt} = \left(r - \frac{2rx}{K}\right)\left(rx\left(1 - \frac{x}{K}\right)\right) = r^2x\left(1 - \frac{2x}{K}\right)\left(1 - \frac{x}{K}\right)$$

and we see that

$$\frac{d^2x}{dt^2} > 0 \text{ if } 0 < x < K/2, \quad \frac{d^2x}{dt^2} < 0 \text{ if } K/2 < x < K$$

so the solutions are the familiar sigmoid shaped logistic curves with a change of concavity at $x = K/2$.

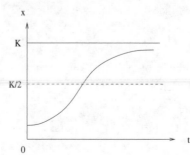

Furthermore, $\left.\frac{dx}{dt}\right|_{x=K/2} = rK/4$ is the maximum growth rate, which we recall is also the critical harvest rate when the population is being constantly harvested. We obtain the plausible conclusion that the population will expire in finite time if the harvest rate exceeds the maximum growth rate.

The example is also important because it beautifully illustrates how just a little simple qualitative analysis can give an excellent description of a solution, without resorting to lengthy analytic or numerical techniques. Merely knowing the equilibrium points $x = 0$, $x = K$, noting that $dx/dt > 0$ for $0 < x < K$, and doing the simple concavity analysis above gives us the logistic curve. Contrast this with the usual example or exercise found in most textbooks to solve the (separable) logistic equation, which ends up finally with an expression like

$$x(t) = \frac{Kx_0}{x_0 + (K - x_0)e^{-rt}}, \quad x(0) = x_0,$$

from which very little insight is immediately evident.

9 Dependence on Initial Conditions

Given the differential equation $\frac{dx}{dt} = f(t, x)$ and two solutions $x_1(t)$, $x_2(t)$ satisfying $x_1(t_0) = a$, $x_2(t_0) = b$ where a and b are close, what can we say about their proximity for values of $t > t_0$? This is the problem of dependence on initial conditions, and to get the needed estimate we will use one of the workhorse tools of nonlinear ODE'ers. Its proof can be found in almost any advanced textbook and is not difficult.

Gronwall's Lemma: If $u(t)$ and $v(t)$ are nonnegative continuous functions on $t_0 \le t < \infty$, and satisfy

$$u(t) \le \alpha + \int_{t_0}^{t} v(s)u(s)\, ds, \qquad t \ge t_0,$$

where α is a nonnegative constant, then

$$u(t) \le \alpha \exp\left[\int_{t_0}^{t} v(s)\, ds \right], \qquad t \ge t_0.$$

We will assume $f(t, x)$ satisfies the conditions for E and U, which implies that there exists a constant $K > 0$ such that the Lipschitz condition,

$$\left| f(t, x) - f(t, y) \right| \le K|x - y|,$$

is satisfied in some neighborhood of (t_0, x_0). The constant K could be an upper bound on $\left| \frac{\partial f}{\partial x} \right|$, for instance. Then each solution satisfies

$$x_i(t) = x_0 + \int_{t_0}^{t} f\big(s, x_i(s)\big)\, ds, \qquad i = 1, 2,$$

where $x_0 = a$ for $x_1(t)$, and $x_0 = b$ for $x_2(t)$, and we are led to the estimate

$$|x_1(t) - x_2(t)| \le |a - b| + \int_{t_0}^{t} \left| f\big(s, x_1(s)\big) - f\big(s, x_2(s)\big) \right| ds.$$

Applying the Lipschitz condition gives

$$|x_1(t) - x_2(t)| \leq |a - b| + \int_{t_0}^{t} K |x_1(s) - x_2(s)| \, ds$$

and then Gronwall's Lemma with $\alpha = |a - b|$, $u(t) = |x_1(t) - x_2(t)|$, and $v(t) = K$, results in the needed estimate

$$|x_1(t) - x_2(t)| \leq |a - b| e^{K(t - t_0)}, \qquad t \geq t_0,$$

which requires some commentary.

First of all, the estimate is a very rough one, but does lead us to the conclusion that if $|a - b|$ is small the solutions will be *close initially*. There is no guarantee they will be intimately linked for t much greater than t_0, nor is the estimate necessarily the best one for a particular case.

Example. If $|a - b| = 10^{-2}, t_0 = 0$, and $K = 2$ then

$$|x_1(t) - x_2(t)| \leq 10^{-2} e^{2t} < \epsilon$$

for $t < \frac{1}{2} \ln(100\epsilon)$; if $\epsilon = \frac{1}{2}$ then $t < 1.956$.

The exponential term in the estimate eventually takes its toll.

The above argument and conclusion is not to be confused with a much emphasized characteristic of chaotic systems—*sensitive dependence on initial conditions*. This means that two solutions starting very close together will rapidly diverge from each other, and later exhibit totally different behavior. The term "rapidly" in no way contradicts the estimate given. If $|a - b| < 10^{-6}$ and $\epsilon = 10^{-4}$ then $t < 2.3025$ in the example above, so "rapidly" requires that at least $t > 2.3025$.

The important fact we can conclude is that solutions are continuous functions of their initial values recognizing this is a local, not global, result.

10   Continuing Continuation

The proof of the E and U Theorem depends on the method of successive approximations, and some fixed point theorem, e.g., the contraction mapping theorem. Given the IVP, $\dot{x} = f(t, x)$, $x(t_0) = x_0$, let $x_0(t) \equiv x_0$, and compute (if possible)

$$x_1(t) = x_0 + \int_{t_0}^{t} f(s, x_0(s)) \, ds, \quad x_2(t) = x_0 + \int_{t_0}^{t} f(s, x_1(s)) \, ds, \ldots,$$

$$x_{n+1}(t) = x_0 + \int_{t_0}^{t} f(s, x_n(s)) \, ds,$$

etc. This is a theoretical construct; after a few iterations, the integrations usually are colossal if not impossible. You want approximations? Use a numerical scheme.

A general set of conditions under which the above successive approximations will converge to a unique solution of the IVP are as follows. First, find a big domain (open, connected set) B in the tx-plane in which $f(t, x)$ and $\partial f(t, x)/\partial x$ are continuous—the proof is the same for a system where $x \in R^n$. Next, construct a closed, bounded box

$\Gamma = \{(t, x) \mid |t - t_0| \leq a, \ |x - x_0| \leq b\}$ contained in B—you can make this as big as you want as long as it stays in B.

Since Γ is closed and bounded, by the properties of continuity we know we can find positive numbers m and k such that $|f(t, x)| \leq m$, $|\partial f(t, x)/\partial x| \leq k$ for (t, x) in Γ. It would be nice if our successive approximations converged to a solution defined in all of Γ, but except for the case where $f(t, x)$ is linear this is usually not the case. To assure convergence we need to find a number $r > 0$ satisfying the three constraints

$$r \leq a, \quad r \leq b/m, \quad \text{and} \quad r < 1/k.$$

Then, the successive approximations $x_n(t)$, $n = 0, 1, 2, \ldots$ converge to a unique solution $x(t)$ of the integral equation

$$x(t) = x_0 + \int_{t_0}^{t} f\big(s, x(s)\big) \, dx,$$

for $|t - t_0| \leq r$. Furthermore, $x(t)$ is a unique solution of the IVP.

But what could happen is that the t-dimension of our rosy-eyed choice of the box Γ may have been considerably reduced to attain an interval on which a solution exists.

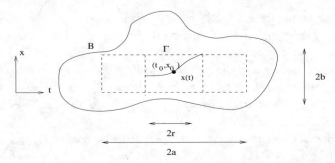

However, looking at the picture above, it appears that we can construct a new box Γ_1 in B with center $(t_0 + r, x(t_0 + r))$, and start a new IVP there. We would create a new value r_1 and by glueing the two solutions together get a solution of our original IVP now defined for $t_0 - r \leq t \leq t_0 + r + r_1$. We could do some carpentry at the other end $(t_0 - r, x(t_0 - r))$ as well.

The continuation theorem says we can continue this process as long as we stay in B, insure that f is bounded in B, and the right- and left-hand limits of the solution exist and are in B as we hit the right- and left-hand boundaries, respectively, of the box Γ. The process continues and the solution is defined for all t, or on some infinite half interval, or it becomes infinite and fails to exist for some finite value of t.

The discussion implies that associated with each initial value (t_0, x_0), and the corresponding solution of the IVP, is a *maximum interval of existence* which is the largest interval on which the solution is defined. It can be thought of as the ultimate finale of the continuation process described above. It is generally impossible to compute except when the IVP can be explicitly solved.

However, the successive approximation scheme applied to a linear equation has one very nice consequence:

Given $\frac{dx}{dt} = a(t)x + b(t)$, where $a(t)$ and $b(t)$ are continuous for $r_1 < t < r_2$, then every solution is defined for $r_1 < t < r_2$.

The result is true for higher dimensional linear systems as well.

Example.

a) $\frac{dx}{dt} = tx + \sin t$, then every solution is defined for $-\infty < t < \infty$.

b) $\frac{dx}{dt} = \frac{1}{t^2-1}x$, then solutions are defined for $-\infty < t < -1$, or $-1 < t < 1$, or $1 < t < \infty$; depending on the choice of t_0 in the IC $x(t_0) = x_0$.

A class of separable equations for which solutions exist on an infinite half line are $\frac{dx}{dt} = f(t)g(x)$, where $f(t)$ and $g(x)$ are continuous,

$$f(t) > 0, t_0 \le t < \infty; g(x) > 0, 0 < x < \infty,$$

and

$$\lim_{x \to \infty} \int_{x_0}^{x} \frac{1}{g(r)} dr = +\infty, \text{ any } x_0 > 0.$$

Then any solution $x(t)$ with IC $x(a) = b$, $a \ge t_0$, $b > 0$, exists on $a \le t < \infty$.

The proof is by contradiction; assume there is a solution $y(t)$ which cannot be continued beyond $t = T$. Since f and g are both positive, $y(t)$ must be strictly increasing, hence $y(t) \to +\infty$ as $t \to T-$. Separate variables to obtain the expression

$$\int_{y_0}^{y(t)} \frac{1}{g(r)} dr = \int_{t_0}^{t} f(s)ds,$$

and as $t \to T-$ the left-hand side becomes infinite while the right-hand side is finite. A simple example of such an equation is $\frac{dx}{dt} = x^\alpha f(t)$, where $0 < \alpha \le 1$, and $f(t)$ is any positive, continuous function.

To see the local nature of the E and U Theorem and the case of the "Incredible Shrinking r" the reader may wish to follow along in this example:

$$\frac{dx}{dt} = 1 + x^2, \quad x(0) = 0.$$

This is the old workhorse with solution $x(t) = \tan t$ defined for $-\pi/2 < t < \pi/2$, and it becomes infinite at the end points. But suppose we don't know this and naively proceed to develop a value of r, then do some continuation.

Since $f(t,x) = 1 + x^2$ is a nice function, and doesn't depend on t, let's choose a box $\Gamma = \{(t,x) \mid |t| \le a, |x| \le b\}$ with center $(0,0)$. Then

$$|f(x)| = 1 + x^2 \le 1 + b^2 = m, \quad \left|\frac{\partial f}{\partial x}\right| = |2x| \le 2b = k$$

and a unique solution is defined for $|t| \le r$ where

$$r = \min\left(a, \frac{b}{1+b^2}, \frac{1}{2b}\right).$$

Since $\max\left(\frac{b}{1+b^2}\right) = \frac{1}{2}$ at $b = 1$, hence $\frac{1}{2b} = \frac{1}{2}$, let $a = \frac{1}{2}$, $b = 1$, then $r = 1/2$, and we would find that $x\left(\frac{1}{2}\right) = \tan(1/2) \cong 0.55$. The solution is defined for $-1/2 \le t \le 1/2$.

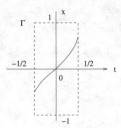

The important point is that no matter how big an $a > 1/2$ and $b > 1$ we initially chose we would get the same 0.55 and 1/2.

At $(1/2, \tan \frac{1}{2}) \approx (1/2, 0.55)$ we can construct a new box $\Gamma_1 = \{(t, x) \mid |t - 1/2| \le a,$ $|x - 0.55| \le b\}$ and get

$$r_1 \le \min \left(a, \frac{b}{(b + 0.55)^2 + 1}, \frac{1}{2(b + 0.55)} \right).$$

An analysis similar to the one above gives $a \approx 0.30$ and $b \approx 1.14$ and $r \approx 0.30$ so our solution has been continued to the right and is guaranteed to exist for $-0.5 \le t \le 0.5 + 0.3 = 0.8$.

Suppose we staggered on and after countless steps our solution reached the point $(1.4, \tan 1.4) \approx (1.4, 5.8)$, and is now defined for $-0.5 \le t \le 1.4$. The box Γ_n would be $\Gamma_n = \{(t, x) \mid |t - 1.4| \le a, |x - 5.8| \le b\}$ and we get

$$r \le \min \left(a, \frac{b}{(b + 5.8)^2 + 1}, \frac{1}{2(b + 5.8)} \right).$$

This gives the miniscule value $r \simeq 0.043$, so we have only continued the solution to the interval $-0.5 \le t \le 1.443$. We expect this since we are precariously close to $\pi/2 \approx 1.57$ where $x(t)$ blows up, so our continuation process will inexorably come to a grinding halt.

3

Insight not Numbers

1 Introduction

The title of this very brief chapter is taken from a quote by the eminent mathematician, Richard Hamming, who said "The purpose of computation is insight not numbers". In dealing with numerical analysis/methods in the introduction to ordinary differential equations this precept should be constantly kept in mind. Too many books nowadays have whole chapters or vast sections discussing a variety of numerical techniques and error estimation, losing sight of the fact that the course is ordinary differential equations not numerical analysis. Examine carefully for the insight issue any book entitled "Differential Equations with ..."—name your favorite mathematical software system.

The emphasis on numerical methods is a little surprising since today's computer software packages all possess very powerful ODE initial value problem solving routines, and even some hand calculators have the capability of approximating solutions to the IVP. This is enhanced by accompanying graphical capabilities, able to do gorgeous direction field or phase plane plots and even 3D visualization. For anyone who long ago labored on step by step Runge–Kutta calculations with pencil, paper and maybe a primitive calculator, or spent a few hours drawing itty bitty arrows to sketch a direction field, today's technology is heaven-sent.

Consequently, the neophyte learner is best served by a brief introduction to some elementary methods to understand how numerical schemes work, some practice using them and the graphics, and later, frequent usage to analyze some meaningful problems or questions. Some discussions have the student develop elaborate programs, forgetting that the course is ordinary differential equations not computer science.

2 Two Simple Schemes

The simplest place to start is with Euler's Method followed by the Improved Euler's Method, sometimes called the Second Order Runge–Kutta or Heun's Method. To approximate the value $x(b)$, $b > a$, of the solution $x(t)$ of the IVP $\dot{x} = f(t, x)$, $x(a) = x_0$, select a step size $h = \frac{b-a}{N}$, N a positive integer, then the two methods are expressed by the simple "do loops":

Euler's Method:

$$t_0 = a$$

$$x_0 = x_0$$

for n from 0 to $N - 1$ do

$$t_{n+1} = t_n + h$$

$$x_{n+1} = x_n + hf(t_n, x_n)$$

print t_N (it better be b!)

print x_N

Improved Euler's Method:

$$t_0 = a$$

$$x_0 = x_0$$

for n from 0 to $N - 1$ do

$$t_{n+1} = t_{n+h}$$

$$u_n = x_n + hf(t_n, x_n)$$

$$x_{n+1} = x_n + \frac{h}{2}\left[f(t_n, x_n) + f(t_{n+1}, u_n)\right]$$

print t_N

print x_N

They are easily understood routines, but it is surprising how some expositors can clutter them up. Modifications can be added such as making a table of all the values $x_n \approx x(t_n)$, $n = 0, \ldots, N$, or plotting them, or comparing the values or plots for different values of the step size h.

3 Key Stuff

What are the important points or ideas to be presented at this time? Here are some:

The Geometry: All numerical schemes essentially are derivative chasing or derivative chasing/correcting/chasing schemes, remembering that we only know *one* true value of $\dot{x}(t)$, that for $t_0 = a$ where $\dot{x}(t_0) = f(t_0, x_0)$. For Euler's Method we use that true value to compute an approximate solution value $x_1 \approx x(t_1)$, then an approximate value of $\dot{x}(t_1) \approx f(t_1, x_1)$, then the next value $x_2 \approx x(t_2)$, etc.

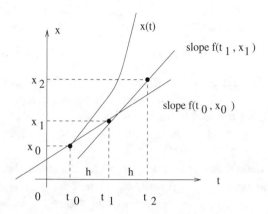

For the Improved Euler Method we compute the approximate value of $\dot{x}(t_1) \simeq f(t_1, x_1)$ using Euler's Method, then use the average of $f(t_0, x_0)$ and $f(t_1, x_1)$ to get a corrected approximation of $x_1 \approx x(t_1)$, then compute the new approximate value of $\dot{x}(t_1) \approx f(t_1, x_1)$, use that to get an approximate value of $\dot{x}(t_2) \approx f(t_2, x_2)$, average again and correct, etc. The picture is:

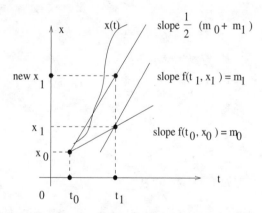

A further nice insight is to note that for the case where the ODE is simply $\dot{x} = f(t)$, $x(a) = x_0$ then $x(t_1) = x_0 + A_1$ where A_1 is the area under the curve $f(t)$, $t_0 \leq t \leq t_1$. Euler's Method uses the left-hand endpoint rectangular approximation to that area, whereas the Improved Euler Method uses the trapezoidal approximation, which is more accurate.

Cost: The Euler Method is a one step method requiring only one function evaluation for each step, whereas the Improved Euler Method requires two. The classical Runge–Kutta Method requires four. For problems where the solution is very wiggly or gets large, and consequently a very small step size h and lots of function evaluations are needed to try to obtain accurate approximations—this comes with a cost. If that cost, which is really a measure of how much number crunching must be done, is too great the little gnome inside the computer might just trash the computation.

Error: The important error of a numerical method is the *global error* which determines the *order* of the method, but almost equally as important is the *local error*. Suppose we are given the IVP $\dot{x} = f(t, x)$, $x(t_0) = x_0$, have selected a step size h, and have generated a sequence of points with the method:

$$x_0 = x(t_0), \quad x_1 \approx x(t_1), \dots, x_n \approx x(t_n), \quad t_{k+1} = t_k + h, \quad k = 0, 1, \dots, n-1.$$

The global error is the value $|x_n - x(t_n)|$, and the method is said to be of *order r* if $|x_n - x(t_n)| = O(h^r)$, which means there is a constant, dependent on things like the initial value and the function f, such that $|x_n - x(t_n)| \leq Mh^r$. For instance the Euler Method is of order 1, the Improved Euler Method is of order 2, and the classic Runge–Kutta is of order 4.

The local error is defined in terms of one step of the iteration as follows: suppose the numerical scheme has landed you at point (t_j, x_j) where $x_j \approx x(t_j)$. For the *new* initial value problem $\dot{\tilde{x}} = f(t, \tilde{x})$, $\tilde{x}(t_j) = x_j$ compute one step of the method to get the point $(t_{j+1}, \tilde{x}_{j+1})$. The local error is the value $|\tilde{x}(t_{j+1}) - \tilde{x}_{j+1}|$. A picture is helpful where ϵ_g is the global error and ϵ_ℓ the local one:

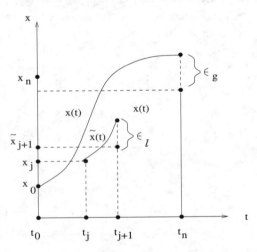

In general, if the global error is of order r the local error is of order $r + 1$.

The important points to be emphasized are

i) If the numerical method is of order r, and we compute with step size h, then again with new step size $\bar{h} = h/m$, where say $m = 2, 4, 10, 10^2, \dots, 10^{23}$ (your choice), we expect the global error to be reduced by $1/m^r$. If $x_N \approx x(t_n)$ for the new step size then

$$|x_n - x(t_n)| \leq Mh^r \quad \text{implies} \quad |x_N - x(t_n)| \leq M\left(\frac{h}{m}\right)^r = \frac{M}{m^r}(h)^r.$$

The last remark has several implications. It is the basis of a reasonable test of the accuracy of the calculations. Compute with step size h, then again with step size h/m where m could equal 2, or 4, or 10, for instance. Decide beforehand the accuracy you need, in terms of significant figures, and compare the two answers. Use their difference to decide whether you need to recompute with a smaller step size or can stop and go have a latté. The second implication is

ii) Do not expect that if you keep reducing the step size you will get closer and closer to the solution. The fiendish roundoff error may contaminate your results; that is usually spotted by heartwarming convergence as the step size is successively reduced, then at the next step the answers go wacky. Also do not expect that even with a very small step size and a reasonable order method you will always get a close approximation. The constant M in the error estimate could be huge, or the method, which computes its own directions, may have gotten off the mark and is merrily computing a solution far removed from the actual one.

4 RKF Methods

This section is not intended as a tutorial on RKF methods (the RK stand for Runge–Kutta, the F for H. Fehlberg who first developed the theory for the method), but to familiarize the reader with the underpinnings of the method. It's not a bad idea to know what the little gnome inside the computer is up to.

The foundation of the method is based on the strategy that by carefully controlling the size of the local error, the global error will also be controlled. Furthermore, a general procedure for estimating the size of the local error is to use two different numerical procedures, with different orders; and compare their difference. Fehlberg was able to do this using Runge–Kutta methods of order 4 and 5, so the local errors are respectively $O(h^5)$ and $O(h^6)$; his method involved six function evaluations at each step.

This was incorporated into a computing scheme which allows for changing the step size h by following this strategy: Selecting a step size h, estimate the relative (to h) size of the local error by using the two method procedure described above; this gives a number $|\text{est.}|$. Next, decide what is an allowable relative local error ϵ, and now use the following criteria in computing (t_{k+1}, x_{k+1}) given (t_k, x_k):

i) If $|\text{est.}| > \epsilon$ reject the x_{k+1} obtained, and compute a new x_{k+1} using a smaller step size.

ii) If $|\text{est.}| < \epsilon$, accept x_{k+1} and compute x_{k+2} using step size h or a larger one.

Of course, in classroom practice the parameters $|\text{est.}|, \epsilon, h$ and others are already built into the RKF45 program so you will see only the computations. Some calculators use an RKF23 program.

The advantages of the RKF45 scheme besides its inherent accuracy are that when the solution is smooth, and the approximations are closely tracking it, the step size selected can be relatively large, which reduces the number of function evaluations (cost!). But if the solution is wiggly the step size can be reduced to better follow it—a fixed step size method could completely overshoot it.

Furthermore, if the solution is too wild or flies off in space the RKF methods have a red flag which goes up if $|\text{est.}| < \epsilon$ cannot be attained. For schemes which allow for adjustments, this requires a bigger value of ϵ be given, otherwise for schemes with preassigned parameters the little gnome will fold its tent and silently steal into the night.

Given the discussion above does the author believe that students should become facile or even moderately familiar with Runge–Kutta methods of order 3 or 4?

Absolutely Not! They require hellishly long tedious calculations or programming because of the number of function evaluations needed to go one step, and why bother when they are available in a computer or calculator package. Use the Euler and Improved Euler Methods to introduce the theory of numerical approximation of solutions of the IVP, and the notion of error. Then give a brief explanation of RKF45, if that is what is available, so they will have an idea of what is going on behind the computer screen. Then develop insight.

5 A Few Examples

The ideal configuration is to have a calculator or computer available to be able to do the Euler and Improved Euler Methods, then have a numerical package like RKF45 to check for accuracy and compute approximations far beyond the capacity of the two methods. Start off with a *few* simple IVPs to gain familiarity with the methods.

> **Comment:** The author finds it hard to support treatments which present a *large* number of problems like:
>
> For the initial value problem $\dot{x} = 2x, x(0) = 1$
>
> a) Use the Euler Method with $h = \frac{1}{4}, \frac{1}{8}$ and $\frac{1}{16}$ to approximate the value of $x(1)$.
>
> b) Use the Improved Euler Method with $h = \frac{1}{4}, \frac{1}{8}$, and $\frac{1}{16}$ to approximate the value of $x(1)$.
>
> c) Find the solution and compute $x(1)$, then construct a table comparing the answers in a) and b).

This is mathematical scut work and will raise the obvious question "If we can find the exact solution why in blazes are we doing this?"

Here are six sample problems—given these as a springboard and the admonitory tone of this chapter the reader should be able to develop many more (insightful) ones. We have used RKF45.

Problem 1 (straightforward). Given the IVP

$$\dot{x} = \sqrt{x + t}, \quad x(0) = 3.$$

Use the Euler Method with $h = 0.05$ and the Improved Euler Method with $h = 0.1$ to approximate $x(1)$ then compare the differences with the answer obtained using RKF45.

Answers: Euler, $h = 0.05$: $x(1) \approx 5.0882078$

Improved Euler, $h = 0.1$: $x(1) \approx 5.1083323$

RKF45: $x(1) \approx 5.1089027$

As expected, Improved Euler gave a much better approximation with twice the step size. The equation can be solved by letting $x = u^2 - t$, but one obtains an implicit solution.

The next three problems use the old veteran $\dot{x} = x^2 + t^2$, whose solutions grow much faster than $\tan t$.

Problem 2. A strategy to estimate accuracy, when only a low order numerical method is in your tool kit, is to compute the solution of an IVP using a step size h, then compute again with step size $h/2$, then compare results for significant figures. Do this for

$$\dot{x} = x^2 + t^2, \quad x(0) = 0$$

and approximate $x(1)$ using a step size $h = 0.2$ and $h = 0.1$ with the Improved Euler Method.

$$\text{Answer: } h = 0.2: \ x(1) \approx 0.356257$$

$$h = 0.1: \ x(1) \approx 0.351830$$

At best, we can say $x(1) \approx 0.35$.

Problem 3. Given the threadbare computing capacity suggested in Problem 2 you can improve your answer using an *interpolation scheme*. Given a second order scheme we know that if $x(T)$ is the exact answer and $x_h(T)$ is the approximate answer using step size h, then

$$x(T) - x_h(T) \approx Mh^2.$$

a) Show this implies that

$$x(T) \approx \frac{4}{3}x_{h/2}(T) - \frac{1}{3}x_h(T)$$

b) Use the result above and the approximations you obtained in Problem 2 to obtain an improved estimate of $x(1)$, then compare it with that obtained using RKF45.

$$\text{Answer: } a) \ \left.\begin{array}{l} x(T) - x_h(T) \approx Mh^2 \\ x(T) - x_{h/2}(T) \approx M\left(\frac{h}{2}\right)^2 = \frac{1}{4}Mh^2 \end{array}\right\} \text{ solve for } x(T)$$

$$b) \ \ x(1) \approx \frac{4}{3}(0.351830) - \frac{1}{3}(0.356257) = 0.350354$$

$$\text{RKF45 gives } x(1) \approx 0.350232$$

Problem 4. For the IVP $\dot{x} = x^2 + t^2$, $x(0) = 0$. Approximate the value of $x(2)$ with the Euler and Improved Euler Methods and step sizes $h = 0.1$ and $h = 0.01$. Compare (graphically if you wish) with the answer obtained with RKF45.

$$\text{Answer: Euler, } h = 0.1: \ x(2) \approx 5.8520996$$

$$\text{Euler, } h = 0.01: \ x(2) \approx 23.392534$$

(Now you suspect something screwy is going on)

$$\text{Improved Euler, } h = 0.1 = x(2) \approx 23.420486$$

(Maybe you start feeling confident)

$$\text{Improved Euler, } h = 0.01: \ x(2) \approx 143.913417 \quad \text{(Whoops!)}$$

$$\text{RKF45: } x(2) \approx 317.724004$$

The awesome power of RKF45 is manifest!

The next problem uses the direction field package of MAPLE, possessed of the ability to weave solutions of initial value problems through the field, to study the existence of periodic solutions of a logistic equation with periodic harvesting.

Problem 5. Given the logistic equation with periodic harvesting

$$\dot{x} = \frac{1}{2}x\left(1 - \frac{x}{4}\right) - \left(\frac{1}{4} + \frac{1}{8}\sin \pi t\right)$$

a) Explain why it has a (stable) 2-periodic ($T = 2$) solution.

b) Such a solution must satisfy $x(0) = x(2)$. Use MAPLE direction field plots and RKF45 to find an approximate value of $x(0)$ then graph the solution. Compute $x(4)$ as a check.

Discussion:

a) Since $\frac{1}{8} \leq \frac{1}{4} + \frac{1}{8}\sin \pi t < \frac{3}{8}$ we see that for $x = 2$ and all t, $\dot{x} = \frac{1}{2} - \left(\frac{1}{4} + \frac{1}{8}\sin \pi t\right) > 0$. For $x = 4$ and all t, $\dot{x} = 0 - \left(\frac{1}{4} + \frac{1}{8}\sin \pi t\right) < 0$, so by an argument given in Ch. 2 there exists a 2-periodic solution $x(t)$ with $2 < x(0) < 4$.

b) Construct a direction field plot for $0 \leq t \leq 2$, $2 \leq x \leq 4$, and select a few initial conditions $x(0) = 2.4$, 2.8, 3.2, 3.6, and plot those solutions. You get something like

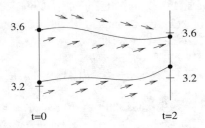

The solution with initial value $x(0) = 3.2$ has terminal value $x(2) > 3.2$, whereas the one with $x(0) = 3.6$ has $x(2) < 3.6$, and it appears that $x(0)$ for the periodic solution is near 3.4. Crank up RFK45: $x(0) = 3.4$, $x(2) = 3.4257$, and now fiddle around or use bisection to get

$$x(0) = 3.454, \quad x(2) = 3.4536, \quad x(4) = 3.4534.$$

Since for $x = 0$ and all t, $\dot{x} = 0 - \left(\frac{1}{4} + \frac{1}{8}\sin \pi t\right) < 0$ there is also an (unstable) periodic solution $x(t)$ with $0 < x(0) < 2$. It will be difficult to approximate because the direction field will veer away from it, but by letting $t \to -t$ and considering the equation

$$\dot{x} = \frac{1}{2}x\left(\frac{x}{4} - 1\right) + \left(\frac{1}{4} + \frac{1}{8}\sin \pi t\right)$$

one can use the previous strategy since now the periodic solution is stable.

Problem 6. Using numerical methods examine the thickness of the boundary layer for the singularly perturbed differential equation

$$\epsilon \dot{x} = -x + (1+t), \quad x(0) = 0, \quad \epsilon = 10^{-1}, 10^{-2}, 10^{-3}.$$

Discussion: From the previous chapter we see that the solution $x(t) \approx 1 + t - \epsilon$ for $t > O(\epsilon)$. The author used a calculator with an RK23 package and produced the following numbers for the IVP:

$$\dot{x} = -\frac{1}{\epsilon}x + \frac{1}{\epsilon}(1+t), \quad x(0) = 0.$$

$\epsilon = 10^{-1}$		$\epsilon = 10^{-2}$		$\epsilon = 10^{-3}$	
t	x	t	x	t	x
0	0	0	0	0	0
0.02	0.1837	0.002	0.1815	0.001	0.6339
0.04	0.3376	0.004	0.3306	0.002	0.8673
0.06	0.4671	0.008	0.5537	0.004	0.9864
0.08	0.5767	0.01	0.6364	0.006	1.0048
0.1	0.6695	0.02	0.8770	0.008	1.0071
0.3	1.1568	0.04	1.0136	0.01	1.0093
0.5	1.3997	0.06	1.0495		
0.7	1.6000	0.08	1.0701		
		0.10	1.0901		

If one takes the thickness of the boundary layers as the first value t_ϵ when $x(t_\epsilon) \approx 1+t-\epsilon$ one sees that they are approximately 0.5, 0.06, 0.006 respectively which supports the $O(\epsilon)$ assertion.

4

Second Order Equations

This chapter could be titled "Second Order Equations—A Second Look," because its intent is not to dwell on the numerous recipes for solving constant coefficient linear equations which stuff the chapters of most introductory texts. Rather, it is to give the conceptual underpinnings for the general second order linear equation, offering some alternative ways of presenting them, recognizing that the discussion is easily generalized to higher order equations. Then a few little morsels are added which enhance some of the standard topics.

1 What Is The Initial Value Problem?

We will write the general second order equation as $\ddot{x} = f(t, x, \dot{x})$, so a solution will be a twice differentiable function $x(t)$ satisfying $\ddot{x}(t) = f(t, x(t), \dot{x}(t))$. But we have only the existence and uniqueness (E and U) theorem for the initial value problem (IVP) at our disposal, so what is the IVP?

Introduce a new dependent variable $y = \dot{x}$ and the equation becomes a two-dimensional first order system

$$\dot{x} = y, \quad \dot{y} = \ddot{x} = f(t, x, \dot{x}) = f(t, x, y).$$

The E and U Theorem in vector form applies: let $\underset{\sim}{x} = \operatorname{col}(x, y)$ then

$$\underset{\sim}{\dot{x}} = \begin{pmatrix} \dot{x} \\ \dot{y} \end{pmatrix} = \begin{pmatrix} y \\ f(t, x, y) \end{pmatrix} = F(t, x, y) = F(t, \underset{\sim}{x})$$

for which the initial conditions are

$$\underset{\sim}{x}(t_0) = \begin{pmatrix} x(t_0) \\ y(t_0) \end{pmatrix} = \underset{\sim}{x}_0 = \begin{pmatrix} x_0 \\ y_0 \end{pmatrix}.$$

We can conclude that the IVP for the second order equation is

$$\ddot{x} = f(t, x, \dot{x}), \quad x(t_0) = x_0, \quad y(t_0) = \dot{x}(t_0) = y_0.$$

The required continuity assumptions for E and U translate to the statement that the IVP will have a unique solution if f, $\partial f/\partial x$, and $\partial f/\partial \dot{x}$ are continuous in some region containing the point (t_0, x_0, y_0).

An easy way to think of this is that we needed one "integration" to solve $\dot{x} = f(t, x)$, so we get one constant of integration to be determined by the initial condition. For the second order equation we need two "integrations" and get two constants. This is easily seen with the simple example

$$\ddot{x} = t^2 \Rightarrow \dot{x}(t) = \frac{t^3}{3} + A \Rightarrow x(t) = \frac{t^4}{12} + At + B,$$

and more abstractly, using the integral equation representation,

$$\dot{x} = y \Rightarrow x(t) = A + \int^t y(s)\,ds$$

and

$$\dot{y} = f(t, x, y) \Rightarrow y(t) = B + \int^t f\big(s, x(s), y(s)\big)\,ds$$

$$= B + \int^t f\left(s, A + \int^s y(r)dr, y(s)\right)\,ds.$$

The above discussion readily generalizes to higher order equations, e.g., for $n = 3$

$$x^{(3)} = f(t, x, \dot{x}, \ddot{x})$$

becomes

$$\dot{x} = y, \quad \dot{y} = z, \quad \dot{z} = f(t, x, y, z).$$

The IVP is

$$x^{(3)} = f(t, x, \dot{x}, \ddot{x}), \quad x(t_0) = x_0, \quad \dot{x}(t_0) = y_0, \quad \ddot{x}(t_0) = z_0,$$

whose solution is a thrice differentiable function $x(t)$ satisfying the initial conditions and the relation

$$x^{(3)}(t) = f\big(t, x(t), \dot{x}(t), \ddot{x}(t)\big)$$

in a neighborhood of (t_0, x_0, y_0, z_0). But we are straying from the hallowed ground of $n = 2$.

IMPORTANT NOTICE

If the following material is to be presented it is incumbent on the presenter to give a few illuminating equations and their solutions, even if the audience isn't quite sure where they came from. Illumination is the stalwart guide to understanding.

2 The Linear Equation

We could first consider the general, n^{th} order, linear, homogeneous equation

$$x^{(n)} + a_1(t)x^{(n-1)} + \cdots + a_{n-1}(t)\dot{x} + a_n(t)x = 0,$$

then later replace the 0 with $q(t)$, a forcing term, to get the nonhomogeneous equation. But we will tenaciously stick with the second order equation, $n = 2$, for several reasons:

a. The theory for the case $n = 2$ carries over easily to higher orders.

b. Higher order linear equations, except possibly for the case $n = 4$ (vibrating beam problems), rarely occur in applications. One reason for this is simply Newton's Law, $F = ma$, in which the acceleration a is a second derivative.

c. Higher order constant coefficient equations require finding the roots of cubic or higher order polynomials. Consequently, unless one wants numerical approximations, most problems and examples are rigged. But factoring polynomials is nearly a lost art, and even the quadratic formula is struggling a little to avoid becoming an endangered species.

Hence sticking with $n = 2$ is a wiser course.

Therefore, we will discuss the IVP

$$(*) \qquad \ddot{x} + a(t)\dot{x} + b(t)x = 0, \quad x(t_0) = x_0, \quad \dot{x}(t_0) = y_0,$$

where $a(t), b(t)$ are continuous for $r < t < s$, and $r < t_0 < s$, and $-\infty < x_0, y_0 < \infty$. Since $f(t, x, \dot{x}) = -b(t)x - a(t)\dot{x}$, then $\frac{\partial f}{\partial x} = -b(t)$, $\partial f/\partial \dot{x} = -a(t)$ are continuous, so the E and U theorem tells us the solution exists and is unique. A deeper and happy fact is that it will be defined for all of $r < t < s$.

Now comes a major fork in the road. One must admit that *exact* solutions of $(*)$ can only be found when

i) $a(t)$ and $b(t)$ are constants,

ii) $(*)$ is an Euler Equation,

$$\ddot{x} + \frac{a}{t}\dot{x} + \frac{b}{t^2}x = 0, \quad t \neq 0, \quad a, b \text{ constant},$$

which is a constant coefficient equation in disguise, or

iii) the good fairy has provided us with a solution $x_1(t)$ and we can use the method of *reduction of order*: assume a second solution is of the form $x_2(t) = x_1(t) \int^t u(s)\, ds$ and substitute in $(*)$:

$$\ddot{x}_2 + a(t)\dot{x}_2 + b(t)x_2 = \left(\ddot{x}_1 \int^t u\, ds + 2\dot{x}_1 u + x_1 \dot{u} \right)$$

$$+ a(t)\left(\dot{x}_1 \int^t u\, ds + x_1 u \right) + b(t)\left(x_1 \int^t u\, ds \right) = 0,$$

since we assumed $x_2(t)$ is a solution. Now combine terms:

$$\left(\ddot{x}_1 + a(t)\dot{x}_1 + b(t)x_1\right) \int^t u\, ds + \left(2\dot{x}_1 u + x_1 \dot{u} + a(t)x_1 u\right) = 0,$$

and since $x_1(t)$ is a solution the first expression is zero and we obtain the first order linear equation for u

$$x_1(t)\dot{u} + \left(2\dot{x}_1(t) + a(t)x_1(t)\right)u = 0.$$

If we can find the solution $u(t)$, and then evaluate its integral, we can find $x_2(t)$, but otherwise the method is pretty much of theoretical value only.

Example. $\ddot{x} + 4t\dot{x} + (4t^2 + 2)x = 0, x_1(t) = e^{-t^2}$. Then $u(t)$ satisfies

$$e^{-t^2}\dot{u} + \left(-4te^{-t^2} + (4t)e^{-t^2}\right)u = e^{-t^2}\dot{u} = 0,$$

hence $u(t) = $ constant, say $u(t) = 1$ and $x_2(t) = e^{-t^2} \int^t ds = te^{-t^2}$.

Remark. In many texts the method of reduction of order assumes $x_2(t) = x_1(t)u(t)$, which results instead in the same first order linear equation, but for $\dot{u}$:

$$x_1(t)\ddot{u} + \left(2\dot{x}_1(t) + a(t)x_1(t)\right)\dot{u} = 0.$$

For all but a few other special equations, one must resort to constructing infinite series solutions, with all the agony of regular points, recurrence relations, regular singular points, indicial equations, logarithmic cases, etc. This is a subject dear to the hearts of special function aficionados, but the neophyte can lose sight of the forest for the indices. More on series, later.

Does this mean one should pay a little obeisance to the notion of linear independence of functions, then go directly to the constant coefficient case where all can be resolved via the quadratic formula and factoring? The author *believes not*, and suggests that a brief sojourn into the general theory of $(*)$ will provide a much stronger foundation and greater understanding.

Two approaches suggest themselves. The first is to deal directly with the structure of the solutions of $(*)$. The second is to discuss the general two-dimensional linear system

$$\dot{x} = a(t)x + b(t)y, \quad \dot{y} = c(t)x + d(t)y$$

and develop briefly the underlying theory, then note that by letting $a(t) = 0$, $b(t) = 1$ we obtain $(*)$ in the form

$$\dot{x} = y, \quad \dot{y} = c(t)x + d(t)y \Rightarrow \ddot{x} - d(t)\dot{x} - c(t)x = 0.$$

This approach is a higher level of exposition, but has the advantage of simultaneously taking care of the two-dimensional system and the important case where $a(t)$, $b(t)$, $c(t)$, and $d(t)$ are all constants. This case is essential to the study of stability of equilibria of almost linear systems. For both approaches the analysis of the nonhomogeneous system, with 0 replaced by $q(t)$ in $(*)$, will be deferred until later in the chapter.

3   Approach 1—Dealing Directly

Linearity is what makes everything tick, and draws on the analogous analysis of algebraic linear systems. Given any solutions $x_1(t)$, $x_2(t), \dots, x_k(t)$ of the second order equation then any linear combination

$$x(t) = \sum_1^k c_j x_j(t)$$

is a solution by rearrangement and linearity of differentiation

$$\ddot{x}(t) + a(t)\dot{x}(t) + b(t)x(t) = \sum_1^k c_j \left(\ddot{x}_j(t) + a(t)\dot{x}_j(t) + b(t)x_j(t) \right)$$

and the last expression is zero since each $x_j(t)$ is a solution.

Linearity will help us answer the question—how many solutions do we need to solve the IVP?

i) If $x_0 = y_0 = 0$ then $x(t) \equiv 0$ is the only solution—this is an important fact.

ii) Given one solution $x(t)$ with $x(t_0) = a \neq 0$, then if $x_0 \neq 0$ the solution $x_1(t) = \frac{x_0}{a}x(t)$ will satisfy $x_1(t_0) = x_0$. But $\dot{x}_1(t_0) = \frac{x_0}{a}\dot{x}(t_0)$ will not equal y_0 unless $\dot{x}(t_0) = \frac{a}{x_0}y_0$, which would be fortuitous. If $y_0 = 0$ then $\dot{x}(t_0)$ must also. A similar analysis can be given for the case $y_0 = b \neq 0$.

Example. $\ddot{x} - 6\dot{x} + 5x = 0$.

A solution is

$$x_1(t) = e^t \quad \text{since} \quad e^t - 6e^t + 5e^t = 0,$$

$$x_2(t) = e^{5t} \quad \text{since} \quad 25e^t - 30e^{5t} + 5e^{5t} = 0.$$

If $t_0 = 0$

$$x_1(t) \quad \text{satisfies} \quad x(0) = 1, \quad \dot{x}(0) = 1$$

$$x_2(t) \quad \text{satisfies} \quad x(0) = 1, \quad \dot{x}(0) = 5.$$

Neither would solve the IVP with $x(0) = 2$, $\dot{x}(0) = -2$, for instance.

The intuitive analysis given above suggests we will need more than one solution.

We can therefore appeal to linearity and ask if given two solutions $x_1(t)$ and $x_2(t)$ can we find constants c_1 and c_2 so that $x(t) = c_1 x_1(t) + c_2 x_2(t)$ satisfies

$$x(t_0) = c_1 x_1(t_0) + c_2 x_2(t_0) = x_0,$$

$$\dot{x}(t_0) = c_1 \dot{x}_1(t_0) + c_2 \dot{x}_2(t_0) = y_0.$$

The constants c_1, c_2 must be unique by uniqueness of the solution of the IVP, and from linear algebra we conclude that $x_1(t), x_2(t)$ must satisfy

$$\det \begin{pmatrix} x_1(t_0) & x_2(t_0) \\ \dot{x}_1(t_0) & \dot{x}_2(t_0) \end{pmatrix} \neq 0.$$

But we have set our sights too low—we want to be able to solve *any* initial value problem for *any* t_0, so the goal is to find two solutions satisfying

$$W(x_1, x_2)(t) = \det \begin{pmatrix} x_1(t) & x_2(t) \\ \dot{x}_1(t) & \dot{x}_2(t) \end{pmatrix} \neq 0, \qquad r < t < s.$$

The determinant is called the *Wronskian*, and we will just denote it by $W(t)$. Observe that

$$\begin{aligned} \frac{d}{dt} W(t) &= \frac{d}{dt}(x_1 \dot{x}_2 - x_2 \dot{x}_1) = \dot{x}_1 \dot{x}_2 + x_1 \ddot{x}_2 - \dot{x}_2 \dot{x}_1 - x_2 \ddot{x}_1 \\ &= x_1(-a(t)\dot{x}_2 - b(t)x_2) - x_2(-a(t)\dot{x}_1 - b(t)x_1) \\ &= -a(t)(x_1 \dot{x}_2 - x_2 \dot{x}_1) = -a(t)W(t); \end{aligned}$$

this is a simple version of Abel's Formula. Its importance is that it implies

$$W(t) = W(t_0) \exp\left[-\int_{t_0}^{t} a(s)ds\right],$$

from which we conclude

> The Wronskian of two solutions is identically zero ($W(t_0) = 0$ for some t_0) or it is never zero.

But if $W(t)$ is never zero that means $x_1(t)$ and $x_2(t)$ can be conjoined to solve any IVP, and we can say $x_1(t)$ and $x_2(t)$ are a *fundamental pair of solutions*.

Before adding more facts, we must ask the fundamental question, begging for an answer.

Does a fundamental pair of solutions exist? Yes, because the solution of the IVP is unique so we can choose

$$x_1(t) \quad \text{satisfying} \quad x_1(t_0) = 1, \quad \dot{x}_1(t_0) = 0,$$

$$x_2(t) \quad \text{satisfying} \quad x_2(t_0) = 0, \quad \dot{x}_2(t_0) = 1.$$

Then $W(t_0) = \det\left(\begin{smallmatrix} 1 & 0 \\ 0 & 1 \end{smallmatrix}\right) = 1 \neq 0$ so $W(t)$ is never zero and the solution of the IVP is

$$x(t) = x_0 x_1(t) + y_0 x_2(t)$$

which can be easily verified. Slick!

But are we stuck with the one fundamental pair of solutions we just found? No, we can find countless others by appealing to a little linear algebra. Let $A = \left(\begin{smallmatrix} a & b \\ c & d \end{smallmatrix}\right)$ be any nonsingular matrix, so $\det A \neq 0$, and given a fundamental pair of solutions $x_1(t)$, $x_2(t)$, then

$$\begin{aligned} \det\left[\begin{pmatrix} x_1(t) & x_2(t) \\ \dot{x}_1(t) & \dot{x}_2(t) \end{pmatrix} A\right] &= \det\begin{pmatrix} ax_1(t) + cx_2(t) & bx_1(t) + dx_2(t) \\ a\dot{x}_1(t) + c\dot{x}_2(t) & b\dot{x}_1(t) + d\dot{x}_2(t) \end{pmatrix} \\ &= \det(W(t)A) = \det W(t) \det A \neq 0. \end{aligned}$$

By linearity the linear combinations $x_3(t) = ax_1(t) + cx_2(t)$, $x_4(t) = bx_1(t) + dx_2(t)$ are solutions, and since their Wronskian is not equal to zero, they are another fundamental pair of solutions. Therefore, the initial value problem can be solved with a linear combination of $x_3(t)$ and $x_4(t)$.

Example. In the previous example $x_1(t) = e^t$, $x_2(t) = e^{5t}$ and

$$W(t) = \det \begin{pmatrix} e^t & e^{5t} \\ e^t & 5e^{5t} \end{pmatrix} = 4e^{6t} \neq 0 \quad \text{for all} \quad t,$$

so $x_1(t)$ and $x_2(t)$ are a fundamental pair. Therefore every solution can be written as

$$x(t) = c_1 e^t + c_2 e^{5t}.$$

If $t_0 = 0$ and the IC are $x(0) = 4$, $\dot{x}(0) = 1$ we would solve $c_1 + c_2 = 4$, $c_1 + 5c_2 = 1$ to obtain $c_1 = 19/4$, $c_2 = -3/4$.

We can find many other fundamental pairs, e.g.

$$x_3(t) = 3e^t - 2e^{5t}, \quad x_4(t) = 7e^t + 3e^{5t}$$

but one simple pair is enough.

The above analysis can be easily carried over to the nth order linear equation, and we can conclude that the space of solutions is an n-dimensional linear space spanned by any set of n fundamental solutions—we have shown this directly for $n = 2$. Any fundamental set of n solutions $x_1(t), x_2(t), \dots, x_n(t)$ will satisfy

$$W(t) = \det \begin{pmatrix} x_1(t) & \cdots & x_n(t) \\ \dot{x}_1(t) & \cdots & \dot{x}_n(t) \\ \vdots & & \vdots \\ x^{(n-1)}(t) & \cdots & x_n^{(n-1)}(t) \end{pmatrix} \neq 0$$

and if the differential equation is

$$x^{(n)} + a_1(t)x^{(n-1)}(t) + \cdots + a_{n-1}(t)\dot{x} + a_n(t)x = 0$$

it can even be shown, using the theory of linear systems, that

$$W(t) = W(t_0) \exp\left[-\int_{t_0}^{t_0} a_1(s)ds\right],$$

hence $W(t)$ is identically zero or never zero. But all the salient facts are covered by limiting the argument to $n = 2$, and a lot of blackboard space is saved as well.

The sharp-eyed knowledgeable reader will surely have noted that the author has avoided the notion of linear independence of functions and solutions. It is there if one appeals to the fact that the column vectors of any nonsingular matrix are linearly independent. For a fundamental pair this means

$$c_1 \begin{pmatrix} x_1(t) \\ \dot{x}_1(t) \end{pmatrix} + c_2 \begin{pmatrix} x_2(t) \\ \dot{x}_2(t) \end{pmatrix} \equiv \begin{pmatrix} 0 \\ 0 \end{pmatrix} \Leftrightarrow c_1 = c_2 = 0,$$

hence

$$c_1 x_1(t) + c_2 x_2(t) \equiv 0 \Leftrightarrow c_1 = c_2 = 0,$$

and this is true for $r < t < s$, which means $x_1(t)$ and $x_2(t)$ are linearly independent.

However, the important distinction is that for a fundamental set of solutions, the Wronskian necessary and sufficient condition applies:

$$W(t) \neq 0 \Leftrightarrow \text{ a fundamental set} \Leftrightarrow \text{ linear independence.}$$

This is not necessarily the case for functions which are not solutions of a linear ODE. The oft quoted example is

$x_1(t) = t^3, x_2(t) = |t|^3$, they are linearly independent on any interval $-r < t < r$, but

$$W(t) = \det \begin{pmatrix} t^3 & |t|^3 \\ 3t^2 & 3t|t| \end{pmatrix} = 0.$$

Both satisfy $x(0) = \dot{x}(0) = 0$ so they could not be solutions of any second order linear equation.

What is being suggested is that the emphasis be put on finding fundamental sets of solutions, which will automatically be linearly independent in view of the Wronskian test, and not bother too much with linear independence, per sé. Too many introductory texts spend excessive time on linear independence, and remember, the subject is ordinary differential equations not algebra.

4 Approach 2—The Linear System

For this treatment all that is needed is a little knowledge of matrix-vector multiplication, which in the 2-dimensional case can be easily learned if needed. We consider the two-dimensional, linear, homogeneous system of differential equations

$$\begin{aligned} \dot{x} &= a(t)x + b(t)y \\ \dot{y} &= c(t)x + d(t)y \end{aligned} \quad \text{or} \quad \begin{pmatrix} \dot{x} \\ \dot{y} \end{pmatrix} = \begin{pmatrix} a(t) & b(t) \\ c(t) & d(t) \end{pmatrix} \begin{pmatrix} x \\ y \end{pmatrix}.$$

If we let $\underset{\sim}{x}$ be the vector $\text{col}(x, y)$ and $A(t)$ be the matrix

$$A(t) = \begin{pmatrix} a(t) & b(t) \\ c(t) & d(t) \end{pmatrix} \quad \text{then} \quad \underset{\sim}{\dot{x}} = A(t)\underset{\sim}{x}$$

and the IVP becomes

$$(\text{IVP}) \quad \begin{cases} \underset{\sim}{\dot{x}} = A(t)\underset{\sim}{x}, \quad \underset{\sim}{x}(t_0) = \underset{\sim}{x_0} = \begin{pmatrix} x_0 \\ y_0 \end{pmatrix}, \\ \qquad\qquad \text{or} \\ \dot{x} = a(t)x + b(t)y, \quad x(t_0) = x_0 \\ \dot{y} = c(t)x + d(t)y, \quad y(t_0) = y_0. \end{cases}$$

If $a(t), b(t), c(t), d(t)$ are continuous on $r < t < s$, and $r < t_0 < s$, then the solution of the IVP will exist and be unique for any $-\infty < x_0, y_0 < \infty$. Since the system is linear we get the added benefit that the solution is defined for all of $r < t < s$.

We will use the matrix $A(t)$ some of the time to avoid writing down the full system, so a solution of the IVP is a vector function $\text{col}(x(t), y(t))$ satisfying

$$\begin{pmatrix} \dot{x}(t) \\ \dot{y}(t) \end{pmatrix} = A(t) \begin{pmatrix} x(t) \\ y(t) \end{pmatrix}, \quad \begin{pmatrix} x(t_0) \\ y(t_0) \end{pmatrix} = \begin{pmatrix} x_0 \\ y_0 \end{pmatrix}.$$

Example.

$$\begin{aligned} \dot{x} &= 2x + y \\ \dot{y} &= 3x + 4y \end{aligned} \quad \text{or} \quad \begin{pmatrix} \dot{x} \\ \dot{y} \end{pmatrix} = \begin{pmatrix} 2 & 1 \\ 3 & 4 \end{pmatrix} \begin{pmatrix} x \\ y \end{pmatrix}.$$

One solution is

$$\underset{\sim}{\varphi_1}(t) = \begin{pmatrix} x(t) \\ y(t) \end{pmatrix} = \begin{pmatrix} e^t \\ -e^t \end{pmatrix} \quad \text{since} \quad \begin{aligned} \frac{d}{dt}(e^t) &= 2(e^t) + (-e^t) \\ \frac{d}{dt}(-e^t) &= 3(e^t) + 4(-e^t) \end{aligned}$$

and if $t_0 = 0$ it satisfies the initial conditions $x(0) = 1$, $y(0) = -1$. Another solution is

$$\underset{\sim}{\varphi_2}(t) = \begin{pmatrix} x(t) \\ y(t) \end{pmatrix} = \begin{pmatrix} e^{5t} \\ 3e^{5t} \end{pmatrix} \quad \text{since} \quad \begin{aligned} \frac{d}{dt}(e^{5t}) &= 2(e^{5t}) + (3e^{5t}) \\ \frac{d}{dt}(3e^{5t}) &= 3(e^{5t}) + 4(3e^{5t}); \end{aligned}$$

it satisfies $x(0) = 1, y(0) = 3$.

We leave the reader to verify linearity: if $\underset{\sim}{\varphi_1}(t), \dots, \underset{\sim}{\varphi_n}(t)$ are solutions then $\underset{\sim}{\varphi}(t) = \sum_1^n c_j \underset{\sim}{\varphi_j}(t)$ is a solution for any constants $c_1, \dots, c_n$. Note that if $\underset{\sim}{\dot{\varphi}_j}(t) = A(t) \underset{\sim}{\varphi_j}$ then

$$\frac{d}{dt}\left(c_j \underset{\sim}{\varphi_j}(t)\right) = c_j \underset{\sim}{\dot{\varphi}_j}(t) = c_j A(t) \underset{\sim}{\varphi_j}(t) = A(t)\left(c_j \underset{\sim}{\varphi_j}(t)\right);$$

interchanging c_j and $A(t)$ is okay since c_j is a scalar.

To satisfy the IVP we need at least two solutions since

$$c \begin{pmatrix} x(t_0) \\ y(t_0) \end{pmatrix} = \begin{pmatrix} x_0 \\ y_0 \end{pmatrix} \quad \text{only if} \quad c = \frac{x_0}{x(t_0)},$$

which then must require that $y_0 = \frac{x_0}{x(t_0)} y(t_0)$. Therefore, given two solutions

$$\underset{\sim}{\varphi_1}(t) = \begin{pmatrix} x_1(t) \\ y_1(t) \end{pmatrix}, \quad \underset{\sim}{\varphi_2}(t) = \begin{pmatrix} x_2(t) \\ y_2(t) \end{pmatrix},$$

we must be able to find two constants c_1, c_2 such that $\underset{\sim}{x}(t) = c_1 \underset{\sim}{\varphi_1}(t) + c_2 \underset{\sim}{\varphi_2}(t)$ satisfies

$$\underset{\sim}{x}(t_0) = \begin{pmatrix} x_0 \\ y_0 \end{pmatrix} \quad \text{or} \quad c_1 \begin{pmatrix} x_1(t_0) \\ y_1(t_0) \end{pmatrix} + c_2 \begin{pmatrix} x_2(t_0) \\ y_2(t_0) \end{pmatrix} = \begin{pmatrix} x_0 \\ y_0 \end{pmatrix}.$$

But this is equivalent to

$$\begin{aligned} c_1 x_1(t_0) + c_2 x_2(t_0) &= x_0 \\ c_1 y_1(t_0) + c_2 y_2(t_0) &= y_0 \end{aligned} \quad \text{or} \quad \begin{pmatrix} x_1(t_0) & x_2(t_0) \\ y_1(t_0) & y_2(t_0) \end{pmatrix} \begin{pmatrix} c_1 \\ c_2 \end{pmatrix} = \begin{pmatrix} x_0 \\ y_0 \end{pmatrix}.$$

From uniqueness of the solution of the IVP, the constants c_1 and c_2 are unique, and since we have the lofty goal of solving the IVP for any choice of t_0, we want to be able to solve the above system of equations for any $t = t_0$.

Consequently we want to find two solutions satisfying

$$W(\varphi_1, \varphi_2)(t) = \det \begin{pmatrix} x_1(t) & x_2(t) \\ y_1(t) & y_2(t) \end{pmatrix} \neq 0.$$

In case the reader skipped over the first approach, the determinant is called the *Wronskian* and we will simply denote it by $W(t)$. Observe that

$$\frac{d}{dt} W(t) = \frac{d}{dt} (x_1 y_2 - x_2 y_1) = \dot{x}_1 y_2 + x_1 \dot{y}_2 - \dot{x}_2 y_1 - x_2 \dot{y}_1,$$

and if one substitutes the expressions for $\dot{x}_1$ and $\dot{y}_1$, $\dot{x}_2$ and $\dot{y}_2$ then combines terms one obtains

$$\frac{d}{dt} W(t) = [a(t) + d(t)] (x_1 y_2 - x_2 y_1) = [a(t) + d(t)] W(t).$$

It follows that

$$W(t) = W(t_0) \exp \left[\int_{t_0}^{t} (a(s) + d(s)) ds \right] = W(t_0) \exp \left[\int_{t_0}^{t} \operatorname{tr} A(s) ds \right]$$

which is Abel's Formula. Recall that the trace of a square matrix A is the sum of its main diagonal terms, and is denoted by $\operatorname{tr} A$.

Therefore, $W(t)$ is identically zero when $W(t_0) = 0$ for some t_0, or it is never zero, in which case a linear combination $c_1 \varphi_1(t) + c_2 \varphi_2(t)$ will solve any IVP. The pair $\varphi_1(t)$, $\varphi_2(t)$ are called a *fundamental pair of solutions*, and we can infer that the solution space of $\dot{x} = A(t)x$ is a linear space of dimension 2.

Example. In the previous example where $A = \begin{pmatrix} 2 & 1 \\ 3 & 4 \end{pmatrix}$ let

$$\varphi_1(t) = \begin{pmatrix} e^t \\ -e^t \end{pmatrix}, \quad \varphi_2(t) = \begin{pmatrix} e^{5t} \\ 3e^{5t} \end{pmatrix}.$$

Then

$$W(t) = \det \begin{pmatrix} e^t & e^{5t} \\ -e^t & 3e^{5t} \end{pmatrix} = 4e^{6t} = 4 \exp \left[\int_0^t \operatorname{tr} A(s) ds \right]$$

so we conclude that they are a fundamental pair, and every solution of $\dot{x} = Ax$ can be written as

$$x(t) = c_1 \begin{pmatrix} e^t \\ -e^t \end{pmatrix} + c_2 \begin{pmatrix} e^{5t} \\ 3e^{5t} \end{pmatrix} = \begin{pmatrix} x(t) \\ y(t) \end{pmatrix}$$

where $x(t) = c_1 e^t + c_2 e^{5t}$, $y(t) = -c_1 e^t + 3c_2 e^{5t}$. If $t_0 = 0$ and the IC were $x(0) = 4$, $y(0) = 1$, we would solve $c_1 + c_2 = 4$, $-c_1 + 3c_2 = 1$ to obtain $c_1 = 11/4$, $c_2 = 5/4$.

To prove the existence of a fundamental pair we proceed as before. Let $\varphi_1(t)$ and $\varphi_2(t)$ be solutions satisfying the initial conditions

$$\varphi_1(t_0) = \begin{pmatrix} x_1(t_0) \\ y_1(t_0) \end{pmatrix} = \begin{pmatrix} 1 \\ 0 \end{pmatrix}, \quad \varphi_2(t_0) = \begin{pmatrix} x_2(t_0) \\ y_2(t_0) \end{pmatrix} = \begin{pmatrix} 0 \\ 1 \end{pmatrix}$$

then $W(t_0) = \det \left(\begin{smallmatrix} 1 & 0 \\ 0 & 1 \end{smallmatrix} \right) = 1 \neq 0$ so $W(t)$ is never zero, hence $\varphi_1(t)$ and $\varphi_2(t)$ are a fundamental pair. The solution of the IVP

$$\dot{x} = A(t)x, \quad x(t_0) = \begin{pmatrix} x_0 \\ y_0 \end{pmatrix} \quad \text{is} \quad x(t) = x_0 \varphi_1(t) + y_0 \varphi_2(t).$$

Countless other pairs of fundamental solutions can be found by noting that given a fundamental pair $\varphi_1(t)$, $\varphi_2(t)$ and any nonsingular matrix $A = \left(\begin{smallmatrix} a & b \\ c & d \end{smallmatrix} \right)$ then

$$\det \left[\begin{pmatrix} x_1(t) & x_2(t) \\ y_1(t) & y_2(t) \end{pmatrix} A \right] = \det \begin{pmatrix} ax_1(t) + cx_2(t) & bx_1(t) + dx_2(t) \\ ay_1(t) + cy_2(t) & by_1(t) + dy_2(t) \end{pmatrix}$$
$$= \det(W(t)A) = \det W(t) \det A \neq 0$$

and since

$$\varphi_3(t) = a \begin{pmatrix} x_1(t) \\ y_1(t) \end{pmatrix} + c \begin{pmatrix} x_2(t) \\ y_2(t) \end{pmatrix}, \quad \varphi_4(t) = b \begin{pmatrix} x_1(t) \\ y_1(t) \end{pmatrix} + d \begin{pmatrix} x_2(t) \\ y_2(t) \end{pmatrix}$$

are solutions by linearity, we have created another fundamental pair.

All of the above analysis carries over to higher dimensional linear systems. The question of linear independence of a set of fundamental solutions is not discussed, and the reader is referred to the final paragraphs of the preceding section, and to linear algebra theory.

Now we can immediately attack the second order equation $\ddot{x}(t) + a(t)\dot{x} + b(t)x = 0$ by noting that it is equivalent to the two-dimensional system

$$\begin{array}{l} \dot{x} = y \\ \dot{y} = -b(t)x - a(t)y \end{array} \quad \text{or} \quad \begin{pmatrix} \dot{x} \\ \dot{y} \end{pmatrix} = \begin{pmatrix} 0 & 1 \\ -b(t) & -a(t) \end{pmatrix} \begin{pmatrix} x \\ y \end{pmatrix}.$$

Then a fundamental pair of solutions

$$\varphi_1(t) = \begin{pmatrix} x_1(t) \\ y_1(t) \end{pmatrix} = \begin{pmatrix} x_1(t) \\ \dot{x}_1(t) \end{pmatrix}, \quad \varphi_2(t) = \begin{pmatrix} x_2(t) \\ y_2(t) \end{pmatrix} = \begin{pmatrix} x_2(t) \\ \dot{x}_2(t) \end{pmatrix},$$

exists for the system, which is equivalent to a fundamental pair of solutions $x_1(t)$, $x_2(t)$ existing for the second order equation. Their Wronskian is

$$W(t) = \det \left(x_1(t)y_2(t) - x_2(t)y_1(t) \right)$$
$$= \det \left(x_1(t)\dot{x}_2(t) - x_2(t)\dot{x}_1(t) \right) = W(t_0) \exp \left[-\int_{t_0}^{t} a(s) \, ds \right]$$

since $\operatorname{tr} A(t) = -a(t)$. Given any fundamental pair $x_1(t)$, $x_2(t)$ another fundamental pair is

$$x_3(t) = ax_1(t) + cx_2(t), \quad x_4(t) = bx_1(t) + dx_2(t),$$

where $ad - bc \neq 0$. Every solution $x(t)$ can be expressed as a linear combination $x(t) = c_1 x_1(t) + c_2 x_2(t)$, where $x_1(t)$, $x_2(t)$ are a fundamental pair.

5 Comments on the Two Approaches

The future presenter of the introduction to the subject may have by now tossed this book on the pile intended for the next library book sale, mumbling about its unsuitability, and returning to the somewhat mind-numbing parade of linear independence, constant coefficient second order equations, constant coefficient higher order equations, constant coefficient linear systems (sometimes with a big dose of linear algebra), etc.

   The problem will be that in a one semester course, this approach will likely leave little time to develop the most important topics, which would certainly include qualitative theory and stability, possibly Hamiltonian systems, the phase plane ($n = 2$!), possibly boundary value problems, and maybe even chaos. If the audience is engineers then knowledge of the Laplace transform is usually required—fortunately there is now good software to avoid the tedious partial fractions. Most of today's introductory texts are replete with mathematical models, many of which are worth exploring, but such explorations benefit from a thorough grounding in the theoretical tools, which are really quite few.

   Time management will be the problem, and the author's suggested approaches—felling two birds with one Wronskian?—may help to relieve the problem, and be able to move faster into the richer plateaus of ordinary differential equations.

6 Constant Coefficient Equations—The Homogeneous Case

At last—we can get our hands on some real solutions! True enough, and it depends on the simple fact that $x(t) = e^{\lambda t}$ is a solution of

$$\ddot{x} + a\dot{x} + bx = 0, \quad a, b \text{ real constants,}$$

if and only if λ is a root of the *characteristic polynomial*

$$p(\lambda) = \lambda^2 + a\lambda + b = 0$$

since $\frac{d^2}{dt^2}(e^{\lambda t}) + a\frac{d}{dt}(e^{\lambda t}) + be^{\lambda t} = e^{\lambda t}p(\lambda)$. The various cases are:

a. $p(\lambda)$ has two distinct real roots $\lambda_1, \lambda_2, \lambda_1 \neq \lambda_2$. Then $x_1(t) = e^{\lambda_1 t}$ and $x_2(t) = e^{\lambda_2 t}$ are solutions and

$$W(t) = \det \begin{pmatrix} e^{\lambda_1 t} & e^{\lambda_2 t} \\ \lambda_1 e^{\lambda_1 t} & \lambda_2 e^{\lambda_2 t} \end{pmatrix} = (\lambda_2 - \lambda_1)e^{(\lambda_1 + \lambda_2)t} \neq 0,$$

so they are a fundamental pair.

b. $p(\lambda)$ has double root λ_1, so $x_1(t) = e^{\lambda_1 t}$ is one solution—what is the other? One can guess $x_2(t) = te^{\lambda_1 t}$ and find out that it works—definitely a time saving approach! But this is a nice time to reintroduce *reduction of order*, an important tool. First note that if λ_1 is a double root of $p(\lambda)$ then from the quadratic formula $\lambda_1 = -\frac{a}{2}$ and $b = \frac{a^2}{4}$.

Now let $x_2(t) = e^{\lambda_1 t} \int^t u(s)ds$ and substitute it into the differential equation to get, after canceling the $e^{\lambda_1 t}$,

$$\dot{u} + (2\lambda_1 + a)u = 0.$$

This implies $\dot{u} = 0$ so $u(t) = A$, and $x_2(t) = Ate^{\lambda_1 t} + Be^{\lambda_1 t}$. The second term is our original $x_1(t)$, and since A is arbitrary we conclude that $x_2(t) = te^{\lambda_1 t}$ is the second solution. Then

$$W(t) = \det \begin{pmatrix} e^{\lambda_1 t} & te^{\lambda_1 t} \\ \lambda_1 e^{\lambda_1 t} & e^{\lambda_1 t} + \lambda_1 te^{\lambda_1 t} \end{pmatrix} = e^{2\lambda_1 t} \neq 0,$$

and therefore $x_1(t)$ and $x_2(t)$ are a fundamental pair.

c. $p(\lambda)$ has complex roots, and since a, b are real they must be complex conjugate pairs

$$\lambda_1 = r + i\theta, \quad \lambda_2 = r - i\theta, \quad \text{where } i^2 = -1.$$

Even if the neophyte has not studied complex variables, it is essential that an important formula be introduced at this point:

$$e^{i\theta} = \cos\theta + i\sin\theta.$$

It can be taken for granted, or motivated by using the infinite series for $\cos\theta$ and $\sin\theta$, and the multiplication rules for products of i. Besides, it leads to the wonderful relation $e^{i\pi} + 1 = 0$, if $\theta = \pi$, wherein $e, i, \pi, 0$ and 1 share in a fascinating concatenation. Letting $\theta \to -\theta$ gives

$$e^{-i\theta} = \cos\theta - i\sin\theta$$

and you obtain the key formulas

$$\cos\theta = \frac{e^{i\theta} + e^{-i\theta}}{2}, \quad \sin\theta = \frac{e^{i\theta} - e^{-i\theta}}{2i}.$$

The law of exponents holds so $x_1(t) = e^{\lambda_1 t} = e^{(r+i\theta)t} = e^{rt}e^{i\theta t}$, and $x_2(t) = e^{\lambda_2 t} = e^{(r-i\theta)t} = e^{rt}e^{-i\theta t}$, are solutions, and any linear combination $x(t) = c_1 x_1(t) + c_2 x_2(t)$ is also a solution. Nothing about linearity prohibits c_1 and c_2 being complex numbers so first let $c_1 = c_2 = \frac{1}{2}$:

$$x_1(t) = \frac{1}{2}e^{rt}e^{i\theta t} + \frac{1}{2}e^{rt}e^{-i\theta t} = e^{rt}\left(\frac{e^{i\theta t} + e^{-i\theta t}}{2}\right) = e^{rt}\cos\theta t,$$

a real solution! Now let $c_1 = \frac{1}{2i}$, $c_2 = -\frac{1}{2i}$ and obtain

$$x_2(t) = \frac{1}{2i}e^{2t}e^{i\theta t} - \frac{1}{2i}e^{rt}e^{-i\theta t} = e^{rt}\left(\frac{e^{i\theta t} - e^{-i\theta t}}{2i}\right) = e^{rt}\sin\theta t,$$

another real solution! Now compute their Wronskian

$$W(t) = \begin{pmatrix} e^{rt}\cos\theta t & e^{rt}\sin\theta t \\ re^{rt}\cos\theta t - \theta e^{rt}\sin\theta t & re^{rt}\sin\theta t + \theta e^{rt}\cos\theta z \end{pmatrix}$$

$$= \theta e^{2rt}(\cos^2\theta t + \sin^2\theta t) = \theta e^{rt} \neq 0$$

since $\theta \neq 0$; they are a fundamental pair.

In these few pages we have covered everything the beginner needs to know about linear, homogeneous (no forcing term), constant coefficient equations, which should be reinforced with some practice. Further practice should come from doing some applied problems.

Remark. The Euler Equation

$$\ddot{x} + \frac{a}{t}\dot{x} + \frac{b}{t^2}x = 0, \qquad t \neq 0,$$

should be briefly mentioned, since it is a camouflaged linear equation and does come up in some partial differential equations, via separation of variables. It also provides absurdly complicated variation of parameters problems for malevolent instructors. Solution technique: let $x(t) = t^\lambda$, substitute to get a characteristic polynomial $p(\lambda) = \lambda^2 + (a-1)\lambda + b$, then work out the three possible cases as above.

Remember, the Euler Equation is one of the few explicitly solvable fly specks in the universe of nonconstant coefficient linear differential equations.

Do you want to venture beyond $n = 2$? You do so at your own risk in view of the fact the majority of high school algebra texts have shelved factoring polynomials in the same musty bin they stored finding square roots. Of course your friendly computer algebra program can do it for you but what is being learned? And then you will *have to* wave your hands and assert the resulting solutions form a fundamental set of solutions because they are linearly independent, unless you want to demonstrate some sleep-inducing analysis.

Example 1. Wish to prove $e^{\lambda_1 t}, e^{\lambda_2 t}, \dots, e^{\lambda_n t}$, λ_i distinct, are linearly independent? You could write down their Wronskian $W(t)$, then

$$W(0) = \det \begin{pmatrix} 1 & \cdots & 1 \\ \lambda_1 & & \lambda_n \\ \vdots & & \vdots \\ \lambda_1^{n-1} & \cdots & \lambda_n^{n-1} \end{pmatrix}$$

which is Vandermonde's determinant (Whoa!) and is never zero if the λ_i are distinct. Or suppose

$$c_1 e^{\lambda_1 t} + c_2 e^{\lambda_2 t} + \cdots + c_n e^{\lambda_n t} = 0, \qquad -\infty < t < \infty,$$

and the c_i are not all zero. Multiply by $e^{-\lambda_1 t}$, differentiate, then multiply by $e^{-(\lambda_1 - \lambda)t}$, differentiate ..., eventually get something like $A c_n e^{(\lambda_n - \lambda_{n-1})t} \equiv 0$ where A depends on all the differences and is not zero so $c_n = 0$. Now repeat to show $c_{n-1} = 0$, etc., making sure to drop a book on the floor in awhile.

Example 2. It gets worse when you have multiple roots λ_j so $e^{\lambda_j t}, te^{\lambda_j t}, \dots,$ $t^k e^{\lambda_j t}$ are solutions if λ_j has multiplicity k. A linear combination of all the solutions being identically zero will produce an expression like

$$p_1(t)e^{\lambda_1 t} + p_2(t)e^{\lambda_2 t} + \cdots + p_n(t)e^{\lambda_n t} = 0, \qquad -\infty < t < \infty,$$

where the $p_i(t)$ are polynomials. Multiply by $e^{-\lambda_1 t}$, and differentiate enough times to make $p_1(t)$ disappear. Now multiply by $e^{-(\lambda_2 - \lambda_1)t}$, differentiate some more, ... Maybe you are a little more convinced of the wisdom of sticking with $n = 2$.

7   What To Do With Solutions

To start with an elementary, but very useful, fact, it is important to be able to construct and use the phase-amplitude representation of a periodic sum:

If $\phi(t) = a\sin\omega t + b\cos\omega t$ then $\phi(t) = A\cos(\omega t - \phi)$ where $A = \sqrt{a^2 + b^2}$
is the *amplitude* and $\phi = \tan^{-1} a/b$ is the *phase angle or phase shift*.

This is trivially proved by multiplying the first expression by $\sqrt{a^2 + b^2}/\sqrt{a^2 + b^2}$ and using some trigonometry. The representation makes for a quick and illuminating graphing technique:

Example. $\phi(t) = 3\sin 2t + 5\cos 2t$ so $A = \sqrt{34}$ and $\phi = \tan^{-1} 3/5 \approx 0.54$ rad., hence $\phi(t) = \sqrt{34}\cos(2t - 0.54)$, and $2t - 0.54 = 0$ gives $t = 0.27$.

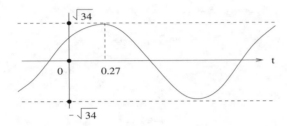

We will shortly see the effective use of the representation in studying stability.

There are three popular models from mechanics used to discuss undamped or damped motion; we first consider the former:

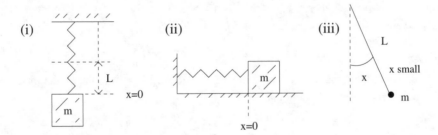

(i) is a suspended mass on a spring, (ii) is the mass connected to a spring moving on a track, and (iii) is a bob or pendulum where x is a small angular displacement from the rest position. In (i) L is the length the spring is stretched when the mass is attached. All share the same equation of motion

$$\ddot{x} + \omega^2 x = 0, \quad \text{the harmonic oscillator,}$$

where $\omega^2 = g/L$ in (i) and (iii), and $\omega^2 = k$, the spring constant in (ii). In the case of the pendulum the equation of motion is an approximation for small displacement to the true equation, $\ddot{x} + \omega^2 \sin x = 0$. Most problems are easy to solve except for the penchant some authors have for mixing units—the length is in meters, the mass is in drams, and it takes place on Jupiter. Stick with the CGS or MKS system!

If there is a displacement x_0 from rest $x = 0$, or an initial velocity y_0 imparted to the mass, then the IVP is

$$\ddot{x} + \omega^2 x = 0, \quad x(0) = x_0, \dot{x}(0) = y_0$$

whose solution is

$$x(t) = \frac{y_0}{\omega} \sin \omega t + x_0 \cos \omega t = \sqrt{x_0^2 + y_0^2/\omega^2} \cos(\omega t - \phi)$$

where $\phi = \tan^{-1} y_0/\omega x_0$. The phase-amplitude expression gives us a hint of the phase plane as follows:

$$\dot{x}(t) = -\omega \sqrt{x_0^2 + y_0^2/\omega^2} \sin(\omega t - \phi)$$

and therefore

$$x(t)^2 + \frac{(\dot{x}(t))^2}{\omega^2} = (x_0^2 + y_0^2/\omega^2).$$

Or if we plot $x(t)$ vs. $\dot{x}(t) = y(t)$ then

$$\frac{x(t)^2}{x_0^2 + y_0^2/\omega^2} + \frac{y(t)^2}{\omega_0^2 x_0^2 + y_0^2} = 1,$$

which is an ellipse. It is the parametric representation of *all* solutions satisfying $x(t_0) = x_0$, $\dot{x}(t_0) = y_0$ for some t_0. Furthermore, we see from the phase-amplitude representation of the solution that a change in the initial time is reflected in a change in the phase angle of the solution, not the amplitude.

If we add damping to the model, which we can assume is due to air resistance or friction (in the case of the pendulum at the pivot), and furthermore make the simple assumption that the dissipation of energy is proportional to the velocity and is independent of the displacement, we obtain the equation of damped motion

$$\ddot{x} + c\dot{x} + \omega^2 x = 0, \qquad c > 0.$$

Now things get more interesting; the standard picture demonstrating the model might add a dashpot to represent the frictional force:

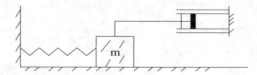

Models of RLC circuits can also be developed which lead to the same equation, but since the author doesn't know an ohm from an oleaster they were omitted.

The motion depends on the nature of the roots of the characteristic polynomial—you could imagine it as the case where the cylinders of your Harley Davidson's shock absorbers were filled with Perrier, Mazola, or Elmer's Glue. The polynomial is $p(\lambda) = \lambda^2 + c\lambda + \omega^2$ with roots

$$\lambda_{1,2} = \frac{1}{2}\left(-c \pm \sqrt{c^2 - 4\omega^2}\right) = \frac{c}{2}\left(-1 \pm \sqrt{1 - \frac{4\omega^2}{c^2}}\right).$$

Case 1. Overdamped, $0 < \frac{4\omega^2}{c^2} < 1$.
Both roots are distinct and negative: $\lambda_1 = -r_1$, $\lambda_2 = -r_2$, and $x(t) = ae^{-r_1t} + be^{-r_2t}$, $r_1, r_2 > 0$, where a and b are determined by the IC. If $r_1 < r_2$ then

$$|x(t)| \le |a|e^{-r_1t} + |b|e^{-r_2t} \le (|a| + |b|)e^{-r_1t}$$

which goes to zero as $t \to \infty$. The solutions will look like this:

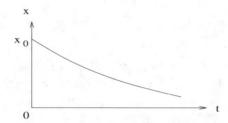

Remark. Most plots depicting this case look like the one above, but the solution can have a small bump before plummeting to zero. Look at $x(t) = 3e^{-t} - 2e^{-2t}$.

Case 2. Critically damped, $\frac{4\omega^2}{c^2} = 1$.
There is one double root $\lambda = -c/2 < 0$ and the solution satisfying $x(0) = x_0$, $\dot{x}(0) = y_0$ is

$$x(t) = x_0e^{-ct/2} + (y_0 + cx_0/2)te^{-ct/2}.$$

If x_0, y_0 are positive it will have a maximum at $t = t_c = \frac{2}{c}\left(\frac{y_0}{y_0+cx_0/2}\right) > 0$ then decrease to zero.

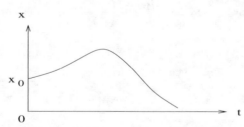

Since the size of the maximum could be of interest in the design of the damping mechanism, assigned problems should be to find the maximum value of $x(t)$ and when it occurs.

Case 3. Underdamped, $\frac{4\omega^2}{c^2} > 1$.
Now the roots are $-\frac{c}{2} \pm i\theta, \theta = \sqrt{\frac{4\omega^2}{c^2} - 1}$, so the solution will be

$$x(t) = ae^{-ct/2}\sin\theta t + be^{-ct/2}\cos\theta t$$

$$= Ae^{-ct/2}\cos(\theta t - \phi)$$

where the maximum amplitude A and the phase angle are determined by the initial conditions. A possible graph would be:

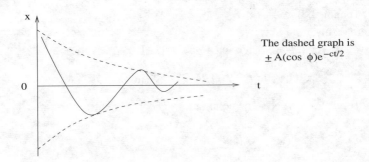

The dashed graph is
$\pm A(\cos \phi)e^{-ct/2}$

The interesting problems for the underdamped case are to estimate a first time beyond which the amplitude is less than a preassigned small value for all later times. These problems afford a nice combination of analysis and the advantages of today's calculator or computer graphing capabilities.

Examples.

1. For the system governed by the IVP

$$\ddot{x} + \frac{1}{8}\dot{x} + x = 0, \quad x(0) = 2, \quad \dot{x}(0) = 0,$$

accurately estimate the smallest value of T for which $|x(t)| < 0.5$ for all $t \geq T$.

Discussion: This can be done solely by using a graphing calculator or a plotting package, but doing some analysis first has benefits. The solution is

$$x(t) \approx 2.0039 e^{-t/16} \cos\left(\frac{\sqrt{255}}{16}t - 0.0626\right),$$

and since $\cos\theta$ has local extrema at $\theta = n\pi$, $n = 0, 1, 2, \ldots$, first solve $\frac{\sqrt{255}}{16}t - 0.0626 = n\pi$ to get $t_n = \frac{16(n\pi + 0.0626)}{\sqrt{255}}$. Then we want

$$|x(t_n)| = 2.0039\exp(-t_n) < 0.5$$

which gives a value of $n \cong 7$. Check some values:

$$n = 7, \quad t_n \approx 22.097, \quad x(t_7) \approx -0.5036$$
$$n = 8, \quad t_n \approx 25.245, \quad x(t_8) \approx 0.4136$$
$$n = 9, \quad t_9 \approx 28.392, \quad x(t_9) \approx -0.3398$$

So we get this picture:

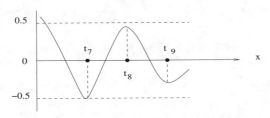

Now call in the number cruncher to solve

$$2.0039e^{-t/16} \cos\left(\frac{\sqrt{255}}{16}t - 0.0629\right) = -0.5$$

with $t_{est} = t_7 \simeq 22.097$ to get $T \approx 22.16968423$ and $x(T) \approx -0.49999999$.

A similar problem is the following:

2. Find a second order linear differential equation and initial conditions whose solution is $x(t) = 2e^{-t/20} \sin t$. Then find an accurate estimate of the smallest value of T for which

$$|x(t)| \le 0.1 \quad \text{for} \quad t \ge T.$$

Discussion: The second half of the problem is like the previous one and $T \approx 58.49217$. But it is surprising how many students have trouble with the first half and go to unreasonable efforts of calculation. From the solution we infer that the roots of the characteristic polynomial are $\lambda_{1,2} = -\frac{1}{20} \pm i$ and since the characteristic polynomial can be written as $\lambda^2 - (\lambda_1 + \lambda_2)\lambda + \lambda_1\lambda_2$ we get the ODE $\ddot{x} + \frac{1}{10}\dot{x} + \left(\frac{1}{400} + 1\right)x = 0$, and from the solution that $x(0) = 0$, $\dot{x}(0) = 2$.

8 The Nonhomogeneous Equation or the Variation of Parameters Formula Untangled

We consider the nonhomogeneous, second order, constant coefficient differential equation

$$(*) \qquad \qquad \ddot{x} + a\dot{x} + bx = q(t),$$

where $q(t)$ is continuous on $r < t < s$, and consequently solutions of the IVP

$$\ddot{x} + a\dot{x} + bx = q(t), \quad x(t_0) = x_0, \quad \dot{x}(t_0) = y_0,$$

will exist, be unique, and be defined for $r < t < s$, any $r < t_0 < s$, and $-\infty < x_0$, $y_0 < \infty$. The goal is to find a *particular solution* $x_p(t)$ of $(*)$, involving no arbitrary constants. Its general solution will then be of the form $x(t) = \phi(t) + x_p(t)$, where $\phi(t)$ is the *general solution* of the homogeneous equation $\ddot{x} + a\dot{x} + bx = 0$.

Since we know a fundamental pair of solutions $x_1(t)$, $x_2(t)$ of the homogeneous equation exists, we can write

$$x(t) = c_1x_1(t) + c_2x_2(t) + x_p(t),$$

then solve for c_1 and c_2 if we are given initial conditions. In some texts, the homogeneous equation is renamed and becomes the complementary equation, and $\phi(t)$ is called the complementary solution. This is old fashioned usage and merely adds to the vocabular baggage.

In the case where $q(t)$ is itself the solution of some constant coefficient, possibly higher order, linear ODE, one can use the Method of Undetermined Coefficients, or Comparison of Coefficients (or Judicious Guessing). The method is sometimes guided by half a page of rules like

If $q(t) = P(t) \cos \omega t + Q(t) \sin \omega t$, where $P(t)$ and $Q(t)$ are polynomials, let k be the larger of the degrees of P and Q. Then a trial particular solution is of the form

$$x(t) = A(t) \cos \omega t + B(t) \sin \omega t,$$

where $A(t)$ and $B(t)$ are polynomials of degree k, with coefficients to be determined by comparison. In the case where $a = 0$ and $b = \omega^2$ the trial particular solution is

$$x_p(t) = t A(t) \cos \omega t + t B(t) \sin \omega t.$$

This may be followed by a detailed table with a column of $q(t)$'s and a parallel column of suggested $x_p(t)$'s. Or the whole method might be justified by the Annihilator Method which gives rules on how to "annihilate" $q(t)$—good material for a mathematical monster movie—"Call the National Guard—the Annihilator has escaped!."

Included with this will be the Principle of Superposition which simply says if $x_{p_i}(t)$ is a particular solution when $q(t) = q_i(t)$, $i = 1, 2, \ldots, k$, then $\sum_1^k x_{p_i}(t)$ is a particular solution when $q(t) = \sum_1^k q_i(t)$. This is a pretty simple consequence of linearity, but its formidable monicker adds more verbal clutter. Following the philosophy that solving equations is not the leitmotif of an introductory course, do a few examples to get a feeling of why the method works and move on.

We come now to the *variation of parameters* method, which was introduced in Chapter 2 to solve first order equations, and as mentioned, the approach taken there goes right over to first order systems and consequently, higher order equations. We want to take that approach because it obviates one major stumbling block present in almost all discussions of the method applied to second order linear equations. We will briefly describe that obstacle.

For the first order equation $\dot{x} = a(t)x + b(t)$, we assumed that a particular solution was of the form $x_p(t) = x(t)u(t)$ where $x(t)$ is a solution of the homogeneous equation and $u(t)$ is an unknown function. Substitute-solve-etc. Since the second order equation (where a and b can be functions of t) has a fundamental pair of solutions $x_1(t)$, $x_2(t)$, to solve the inhomogeneous equation

$$\ddot{x} + a(t)\dot{x} + b(t)x = q(t)$$

it makes sense to assume a particular solution of the form

$$x_p(t) = x_1(t)u_1(t) + x_2(t)u_2(t).$$

Now substitute and try to solve for $u_1(t)$ and $u_2(t)$, and here is where the problem arises. In computing $\dot{x}_p(t)$ one obtains after regrouping

$$\dot{x}_p = \dot{x}_1 u_1 + \dot{x}_2 u + (x_1 \dot{u}_1 + x_2 \dot{u}_2)$$

and to successfully complete the argument one must "impose" the condition that the parenthetical expression equals zero:

$$x_1(t)\dot{u}_1(t) + x_2(t)\dot{u}_2(t) = 0.$$

The arguments found in most textbooks to justify this are pretty hazy: some examples (paraphrased) are:

We might expect many possible choices of u_1 and u_2 to meet our need.

We may be able to impose a second condition of our own choice to make the computation more efficient.

To avoid the appearance of second derivatives (of the u_i) we are free to impose an additional condition of our own choice.

(Okay, I decided that I want $x_1\dot{u}_1 + x_2\dot{u}_2 = 13.4079$—now what?). The authors should *not* be faulted for using a little legerdemain, since they are partly victims of the traditional path they have followed to get to this point.

A much more enlightening approach is to mimic the variation of parameters derivation used previously for first order equations. Recall that for the equation $\dot{x} = a(t)x + b(t)$ one assumes a particular solution of the form $x_p(t) = \Phi(t)u(t)$, where $\Phi(t) = \exp\left[\int^t a(s)\,ds\right]$. Then it is substituted into the ODE to obtain a first order differential equation, which is solved to get

$$x_p(t) = \Phi(t)\int \Phi^{-1}(s)b(s)\,ds.$$

But we can apply the same idea to the nonhomogeneous system

$$\dot{x} = a(t)x + b(t)y + p(t), \quad \dot{y} = c(t)x + d(t)y + q(t)$$

if we first rewrite it as

$$\dot{\underset{\sim}{x}} = \begin{pmatrix} \dot{x} \\ \dot{y} \end{pmatrix} = A(t)\underset{\sim}{x} + B(t) = \begin{pmatrix} a(t) & b(t) \\ c(t) & d(t) \end{pmatrix}\begin{pmatrix} x \\ y \end{pmatrix} + \begin{pmatrix} p(t) \\ q(t) \end{pmatrix},$$

then determine what is $\Phi(t)$, a fundamental matrix. Our previous discussion has produced the solution of the homogeneous system

$$\underset{\sim}{x}(t) = \begin{pmatrix} x(t) \\ y(t) \end{pmatrix} = \begin{pmatrix} x_1(t) & y_1(t) \\ x_2(t) & y_2(t) \end{pmatrix}\begin{pmatrix} x_0 \\ y_0 \end{pmatrix}$$

where $\mathrm{col}(x_1(t), x_2(t))$, $\mathrm{col}(y_1(t), y_2(t))$ are a pair of fundamental solutions and there's our $\Phi(t)$! Furthermore it is invertible since it is nonsingular and we can always choose $x_1(t)$, $x_2(t)$ so that $\Phi(t_0) = I$.

Now just mimic what was done before, first noting that $\Phi(t)$ satisfies the matrix ODE

$$\dot{\Phi} = \begin{pmatrix} \dot{x}_1 & \dot{y}_1 \\ \dot{x}_2 & \dot{y}_2 \end{pmatrix} = \begin{pmatrix} a(t) & b(t) \\ c(t) & d(t) \end{pmatrix}\begin{pmatrix} x_1 & y_1 \\ x_2 & y_2 \end{pmatrix} = A(t)\Phi.$$

The variation of parameters assumption is that the particular solution $\underset{\sim}{x}_p(t)$ can be expressed as $\underset{\sim}{x}_p(t) = \Phi(t)\underset{\sim}{u}(t)$, so substitute:

$$\dot{\underset{\sim}{x}}_p = \dot{\Phi}\underset{\sim}{u} + \Phi\dot{\underset{\sim}{u}} = A(t)\Phi\underset{\sim}{u} + B(t).$$

Since $\dot{\Phi} = A(t)\Phi$, the first term in each expression cancels and we get

$$\Phi\dot{\underset{\sim}{u}} = B(t) \Rightarrow \dot{\underset{\sim}{u}} = \Phi^{-1}(t)B(t) \Rightarrow \underset{\sim}{u}(t) = \int^t \Phi^{-1}(s)\underset{\sim}{B}(s)\,ds$$

hence

$$x_p(t) = \Phi(t) \int^t \Phi^{-1}(s) \underset{\sim}{B}(s) \, ds.$$

If we are solving the IVP, we could put limits on the integral and use $\int_{t_0}^t$, then $\underset{\sim}{x}_p(t_0) = 0$, so the particular solution plays no role in the evaluation of the necessary constants in the general solution.

But we were sticking to $n = 2$ weren't we? Yes, and that makes life even easier since inverting 2×2 matrices is a piece of cake. Recalling that

$$\det \begin{pmatrix} x_1(t) & y_1(t) \\ x_2(t) & y_2(t) \end{pmatrix} = W(t) \neq 0,$$

we have

$$\Phi^{-1}(t) = \begin{pmatrix} x_1(t) & y_1(t) \\ x_2(t) & y_2(t) \end{pmatrix}^{-1} = \frac{1}{W(t)} \begin{pmatrix} y_2(t) & -y_1(t) \\ -x_2(t) & x_1(t) \end{pmatrix}$$

and now we can write down the expression for the particular solution in all its splendor— but we won't. The topic of this chapter is second order equations, and we don't want to stray too far from the terra firma of $n = 2$.

So consider the special case

$$A(t) = \begin{pmatrix} 0 & 1 \\ -b(t) & -a(t) \end{pmatrix}, \quad \underset{\sim}{B}(t) = \begin{pmatrix} 0 \\ q(t) \end{pmatrix}$$

which corresponds to the second order equation

$$\ddot{x} + a(t)\dot{x} + b(t)x = q(t).$$

In this case given a fundamental pair of solutions $x_1(t)$, $x_2(t)$, let

$$\Phi(t) = \begin{pmatrix} x_1(t) & x_2(t) \\ \dot{x}_1(t) & \dot{x}_2(t) \end{pmatrix}, \quad \Phi^{-1}(t) = \frac{1}{W(t)} \begin{pmatrix} \dot{x}_2(t) & -x_2(t) \\ -\dot{x}_1(t) & x_1(t) \end{pmatrix}$$

and $W(t) = \exp\left[-\int^t a(s)\,ds\right]$, therefore

$$\underset{\sim}{x}_p(t) = \begin{pmatrix} x_p(t) \\ \dot{x}_p(t) \end{pmatrix} = \begin{pmatrix} x_1(t) & x_2(t) \\ \dot{x}_1(t) & \dot{x}_2(t) \end{pmatrix} \int^t \frac{1}{W(s)} \begin{pmatrix} \dot{x}_2(s) & -x_2(s) \\ -\dot{x}_1(s) & x_1(s) \end{pmatrix} \begin{pmatrix} 0 \\ q(s) \end{pmatrix} ds.$$

But we are really only interested in the expression for $x_p(t)$, so after a little matrix multiplication we pick out the first component of $\underset{\sim}{x}_p(t)$ and get

$$x_p(t) = -x_1(t) \int^t \frac{x_2(s)q(s)}{W(s)} \, ds + x_2(t) \int^t \frac{x_1(s)q(s)}{W(s)} \, ds.$$

This is exactly what we would get if we solved the system of equations for $\mathrm{col}(\dot{u}_1, \dot{u}_2)$,

$$x_1\dot{u}_1 + x_2\dot{u}_2 = 0, \quad \dot{x}_1\dot{u}_1 + \dot{x}_2\dot{u}_2 = q(t),$$

then integrated. Note these contain the "mysterious" first equation which never appears in the systems approach.

For the constant coefficient case we will derive an alternate, and often more useful, form of the particular solution, but we will not expand the discussion to higher order equations or to systems. Just keep the encapsulating statement in mind:

The general solution of the system $\dot{\underset{\sim}{x}} = A(t)\underset{\sim}{x} + \underset{\sim}{B}(t)$ is given by the expression

$$\underset{\sim}{x}(t) = \Phi(t)\underset{\sim}{c} + \Phi(t) \int^t \Phi^{-1}(s)\underset{\sim}{B}(s)\, ds$$

where $\Phi(t)$ is any fundamental matrix of the homogeneous system $\dot{\underset{\sim}{x}} = A(t)\underset{\sim}{x}$ and $\underset{\sim}{c}$ is determined by initial conditions, if any.

Note that the expression for $\underset{\sim}{x}_p(t)$ is independent of the choice of a fundamental matrix $\Phi(t)$, since if $\Phi_1(t) = \Phi(t)Q$, where $\det Q \neq 0$ then

$$\Phi_1(t) \int^t \Phi_1^{-1}(s)\underset{\sim}{B}(s)\, ds = \Phi(t)Q \int^t Q^{-1}\Phi^{-1}(s)\underset{\sim}{B}(s)\, ds = \Phi(t) \int^t \Phi^{-1}(s)\underset{\sim}{B}(s)\, ds.$$

Example. $\ddot{x} + \frac{1}{4t^2}x = q(t), t \neq 0$, an Euler equation with $x_1(t) = t^{1/2}$, $x_2(t) = t^{1/2}\log t$, and since $a(t) = 0$, $W(t) = 1$.

$$\Phi(t) = \begin{pmatrix} t^{1/2} & t^{1/2}\log t \\ t^{-1/2}/2 & t^{-1/2} + \frac{1}{2}t^{-1/2}\log t \end{pmatrix}$$

so

$$\Phi^{-1}(t) = \begin{pmatrix} t^{-1/2} + \frac{1}{2}t^{-1/2}\log t & -t^{1/2}\log t \\ -t^{-1/2}/2 & t^{1/2} \end{pmatrix}$$

and therefore

$$\begin{pmatrix} x_p(t) \\ \dot{x}_p(t) \end{pmatrix} = \begin{pmatrix} t^{1/2} & t^{1/2}\log t \\ t^{-1/2}/2 & t^{-1/2} + \frac{1}{2}t^{-1/2}\log t \end{pmatrix}$$
$$\cdot \int^t \begin{pmatrix} s^{-1/2} + \frac{1}{2}s^{-1/2}\log s & -s^{1/2}\log s \\ -s^{-1/2}/2 & s^{1/2} \end{pmatrix} \begin{pmatrix} 0 \\ q(s) \end{pmatrix} ds$$

from which we obtain the expression

$$x_p(t) = -t^{1/2} \int^t (s^{1/2}\log s)q(s)\, ds + t^{1/2}\log t \int s^{1/2}q(s)\, ds.$$

The example points out the real problem with variation of parameters—any exercises are not much more than tests of integration skill, e.g. let $q(t) = t^{3/2}$ above. More likely, the answers can only be computed numerically, e.g. let $q(t) = \cos t$ above, or even simpler, solve $\ddot{x} + x = t^{1/2}$. The importance of the variation of parameters method is that, for the system, it gives an exact integral representation of the solution, which can be very useful in determining bounds on solutions, long time behavior, and stability if we have information about $\Phi(t)$ and $B(t)$. We will see examples of this in Chapter 5.

The best advice, whatever method or approach one uses to introduce variation of parameters, is to work a few problems then stick a bookmark in that section of the book in case you ever need to use the formula.

9 Cosines and Convolutions

The most interesting problems for the second order, constant coefficient equation are those generally described as the forced (damped or undamped) oscillator. A typical mechanical configuration is a spring/mass system being driven by a periodic forcing term:

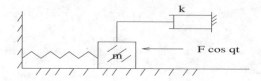

For simplicity we will assume that the natural frequency (when k and F are zero) of the system is $\omega = 1$, so the equation describing the system is

$$\ddot{x} + k\dot{x} + x = F\cos qt, \qquad k > 0.$$

Solving the equation gives a solution of the form

$$x(t) = x_u(t) + K\cos(qt - \phi)$$

where

$\qquad x_u(t)$ is the solution of the unforced system $(F = 0)$
$\qquad K\cos(qt - \phi)$ is the particular solution.

Since $k > 0$ we know $x_u(t) \to 0$ as $t \to \infty$; it is called the *transient solution*. Hence $x(t) \to K\cos(qt - \phi)$, the *steady state solution*. But all is not as simple as it appears, so we need to look at some cases.

Case 1: damping: $k > 0$, and $q = 1$.
The frequency of the forcing term matches the natural frequency. Solving the equation we have

$$x(t) = x_u(t) + \frac{F}{k}\cos(t - \pi/2) = x_u(t) + \frac{F}{k}\sin t,$$

so the solution approaches a natural mode of the undamped system $\ddot{x} + x = 0$. Note that the amplitude F/k can be very large if $k > 0$ is small.

Case 2: damping: $k > 0$, and $q \neq 1$.
A little calculation results in the steady state solution

$$x(t) = FK(q)\cos(qt - \phi), \quad K(q) = \frac{1}{\sqrt{(1 - q^2)^2 + k^2 q^2}},$$

and $K(q)$, which multiplies F, the amplitude of the forcing term, is called the *amplification factor or gain*.

It is a simple first semester calculus exercise to show that

i) $K(0) = 1, K(q) \to 0$ as $q \to \infty$.

ii) $K'(q) = 0$ for $q^2 = 1 - \frac{1}{2}k^2$, which will be positive when $k < \sqrt{2}$, and in this case it will have a maximum $(k\sqrt{1 - k^2/4})^{-1}$. Otherwise $(k > \sqrt{2})$ and $K(q)$ monotonically decreases to zero.

iii) If $k = 0$, $K(q)$ will have a vertical asymptote at $q = 1$.

A graph of the gain vs. q looks like

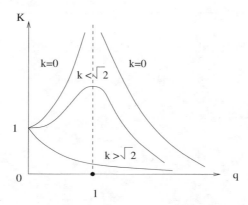

If we fix k at a value $1 < k < \sqrt{2}$ we can "tune" the system to get maximum amplitude by adjusting q, the forcing frequency. A good practical problem could be: let $F = 10$ then plot $K(q)$ for various values of k, and compute the maximum gain, if any, and the phase shift in each case.

The Undamped Cases: $k = 0$

The equation is $\ddot{x} + x = F\cos qt$ and if we assume the system starts at rest so $x(0) = \dot{x}(0) = 0$, there are two cases

Case 1: $q \neq 1$

The solution is $x(t) = \frac{F}{1-q^2}(\cos qt - \cos t)$ a superposition of two harmonics, with a large amplitude if q is close to 1. Via a trigonometric identity for the difference of two cosines we get

$$x(t) = \frac{2F}{1 - q^2}(\sin(q + 1)t)(\sin(1 - q)t)$$

and suppose q is close to 1, say $q = 1 + \epsilon$, $\epsilon > 0$. Then

$$x(t) = \left[\left(\frac{2F}{2\epsilon + \epsilon^2}\right)\sin \epsilon t\right] \sin(2 + \epsilon)t$$

and the bracketed term will have a large amplitude and a large period $2\pi/\epsilon$. The other term will have a period $\frac{2\pi}{2+\epsilon} \sim \pi$, so we see the phenomenon of *beats*, where the large amplitude, large period oscillation is an envelope which encloses the smaller amplitude, smaller period oscillation.

Case 2: $q = 1$

Instead of using undetermined coefficients, or variation of parameters for the intrepid souls, to solve this case, go back to the first expression for the solution in the case $q \neq 1$, and let $q = 1 + \epsilon$. We have

$$x(t) = \frac{F}{1 - (1 + \epsilon)^2} [\cos(1 + \epsilon)t - \cos t]$$

and with a little trigonometry and some readjustment this becomes

$$x_\epsilon(t) = \frac{-F}{2 + \epsilon} \left[(\cos t)\frac{\cos \epsilon t - 1}{\epsilon} - (\sin t)\frac{\sin \epsilon t}{\epsilon} \right].$$

Now use L'Hôpital's Rule letting $\epsilon \to 0$ to get

$$\lim_{\epsilon \to 0} x_\epsilon(t) = x(t) = \frac{F}{2}t \sin t,$$

which is the dreaded case of *resonance*.

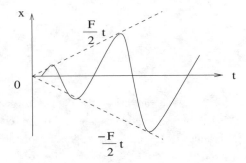

For the more general case $\ddot{x} + \omega^2 x = F \cos \omega t$ we get $x(t) = \frac{F}{2\omega}t \sin \omega t$. A sample problem might be one similar to those suggested for the underdamped system:

A system governed by the IVP

$$\ddot{x} + 4x = \frac{16}{\pi} \cos 2t, \quad x(0) = \dot{x}(0) = 0,$$

"blows up" when the amplitude $|x(t)|$ exceeds 24. When does this first occur?

A more general discussion of resonance can be obtained for the constant coefficient case by slightly adjusting the form of the variation of parameters formula, or using the following result:

The solution $x(t)$ of the IVP

$$\ddot{x} + a\dot{x} + bx = q(t), \quad x(0) = \dot{x}(0) = 0$$

is given by the convolution integral

$$x_p(t) = \int_0^t \phi(t - s)q(s)\,ds$$

where $\phi(t)$ is the solution of the homogeneous differential equation satisfying $\phi(0) = 0$, $\dot{\phi}(0) = 1$.

To prove the result, first compute the derivatives of $x_p(t)$:

$$\dot{x}_p(t) = \phi(t-t)q(t) + \int_0^t \dot{\phi}(t-s)q(s)\,ds = \int_0^t \dot{\phi}(t-s)q(s)\,ds \qquad \text{since } \phi(0) = 0,$$

$$\ddot{x}_p(t) = \dot{\phi}(t-t)q(t) + \int_0^t \ddot{\phi}(t-s)q(s)\,ds = q(t) + \int_0^t \ddot{\phi}(t-s)q(s)\,ds \text{ since } \dot{\phi}(0) = 1.$$

Then

$$\ddot{x}_p + a\dot{x}_p + bx_p = q(t) + \int_0^t (\ddot{\phi} + a\dot{\phi} + b\phi)(t-s)q(s)\,ds = q(t)$$

since $\phi(t)$ is a solution of the homogeneous equation.

A more elaborate proof would be to use the variation of parameters formula for each of the constant coefficient cases and combine terms.

The convolution integral representation has several very nice features:

a. Since $x_p(0) = \dot{x}_p(0) = 0$, the particular solution contributes nothing to the initial conditions, which makes the IVP

$$\ddot{x} + a\dot{x} + bx = q(t), \quad x(0) = x_0, \quad \dot{x}(0) = y_0$$

easier to solve. One merely adds to $x_p(t)$ the solution of the homogeneous problem satisfying the IC.

b. Similar to the case for the first order equation it gives an elegant representation of the solution when $q(t)$ is a discontinuous function.

Example. $\ddot{x} - 2\dot{x} + x = \begin{cases} -1, & 0 \le t \le 2 \\ 1, & 2 < t \end{cases}$. Then $\phi(t) = te^t$ and

$$x_p(t) = \int_0^t (t-s)e^{t-s}(-1)\,ds, \qquad 0 \le t \le 2,$$

$$x_p(t) = \int_0^2 (t-s)e^{t-s}(-1)\,ds + \int_2^t (t-s)e^{t-s}(1)\,ds, \qquad t > 2.$$

Working out the integrals gives

$$x_p(t) = -te^t + e^t - 1, \qquad 0 \le t \le 2,$$
$$x_p(t) = te^{t-2} - 3e^{t-2} - e^2, \qquad 2 < t.$$

But the real value of the convolution integral representation is that it gives a very specific criteria for when resonance can occur in the general case. Consider the IVP

$$\ddot{x} + x = f(t), \quad x(0) = x_0, \quad \dot{x}(0) = y_0,$$

where $f(t + 2\pi) = f(t)$. In this case $\phi(t) = \sin t$ so

$$x(t) = x_0 \cos t + y_0 \sin t + x_p(t),$$

$$x_p(t) = \int_0^t \sin(t-s)f(s)\,ds.$$

Then

$$x_p(t + 2\pi) = \int_0^{t+2\pi} \sin(t - s)f(s)\,ds$$

$$= \int_0^t \sin(t - s)f(s)\,ds + \int_t^{t+2\pi} \sin(t - s)f(s)\,ds$$

and by making a change of variables in the second integral and since $f(t + 2\pi) = f(t)$ we get

$$x_p(t + 2\pi) = x_p(t) + \int_0^{2\pi} \sin(t - s)f(s)\,ds.$$

It follows from the last equation that

$$x_p(t + 4\pi) = x_p(t + 2\pi) + \int_0^{2\pi} \sin(t + 2\pi - s)f(s)\,ds$$

$$= x_p(t) + 2\int_0^{2\pi} \sin(t - s)f(s)\,ds,$$

and in general for any positive integer N

$$x_p(t + 2N\pi) = x_p(t) + N\int_0^{2\pi} \sin(t - s)f(s)\,ds.$$

But if there is some value $t = \alpha$ having the property that $\int_0^{2\pi} \sin(\alpha - s)f(s)\,ds = m \neq 0$, then

$$x_p(\alpha + 2N\pi) = x_p(\alpha) + Nm$$

which becomes unbounded as $N \to \infty$. Resonance!

We conclude that

The equation $\ddot{x} + x = f(t)$, $f(t + 2\pi) = f(t)$, will have a 2π-periodic solution whenever $\int_0^{2\pi} \sin(t-s)f(s)\,ds = 0$ for all t. Otherwise, there will be resonance.

The result allows us to consider a variety of functions $f(t)$ satisfying $f(t + 2\pi) = f(t)$, for example

a) $f(t) = |\sin t|$, then

$$\int_0^{2\pi} (\sin(t - s))|\sin s|\,ds = \int_0^{\pi} \sin(t - s)\sin s\,ds$$

$$+ \int_\pi^{2\pi} \sin(t - s)(-\sin s)\,ds = 0,$$

so no resonance

b) $f(t) = \begin{cases} -1 & 0 < t < \pi \\ 1 & \pi < t < 2\pi \end{cases}$ periodically extended, then

$$\int_0^{2\pi} \sin(t-s)f(s)\,ds = -\int_0^{\pi} \sin(t-s)\,ds$$

$$+ \int_{\pi}^{2\pi} \sin(t-s)\,ds = 4\cos t$$

so resonance occurs.

The above result is a specific instance of a more general result for constant coefficient systems $\dot{\underset{\sim}{x}} = A\underset{\sim}{x}$. It states that if $\Phi(t)$ is the fundamental matrix satisfying $\Phi(0) = I$, the identity matrix, then $\Phi(t-s) = \Phi(t)\Phi^{-1}(s)$ for all t and s. Note furthermore that $\int_0^{2\pi} \sin(t-s)f(s)\,ds = 0$ for all t is equivalent to

$$\sin t \int_0^{2\pi} \cos s\; f(s)\,ds - \cos t \int_0^{2\pi} \sin s\; f(s)\,ds = 0, \text{ for all } t.$$

In particular, this would be true for $t = \pi/2$ and $t = \pi$ which implies that resonance will not occur if the orthogonality relations

$$\int_0^{2\pi} \cos s\, f(s)\,ds = 0, \qquad \int_0^{2\pi} \sin s\, f(s)\,ds = 0$$

are satisfied.

10 Second Thoughts

There are some other facets of second order, linear, differential equations which could be part of the introductory course, and they merit some comment.

Numerical Methods

For second order equations numerical methods have no real place in a first course, assuming that the usual (hopefully not too long) introduction to the topic was given when first order equations were discussed. What should be made clear is that despite what the computer screen might show, the numerical schemes convert the second order equation to a first order system—remember they can only chase derivatives. For example, to use Euler's Method for the general IVP

$$\ddot{x} = f(t, x, \dot{x}), \quad x(a) = x_0, \quad \dot{x}(a) = y_0,$$

it is first converted to a first order system

$$\dot{x} = y, \quad \dot{y} = f(t, x, y), \quad x(a) = x_0, \quad y(a) = y_0.$$

Then the algorithm to approximate the values $x(b)$, $y(b) = \dot{x}(b)$ is:

Euler's Method

$$t_0 = a$$

$$x_0 = x_0$$

$$y_0 = y_0$$

for n from 0 to $N - 1$ do

$$t_{n+1} = t_n + h$$

$$x_{n+1} = x_n + hy_n$$

$$y_{n+1} = y_n + hf(t_n, x_n, y_n)$$

Print t_N, x_N, y_N

where $h = \frac{b-a}{N}$.

Once this is explained, leave the numerics to the little gnome and enjoy his sagacity.

Boundary Value Problems   Unless the presentation is intended as preparation for a course in partial differential equations, and consequently the emphasis is on linear differential equations, there will be little time for boundary value problems. The Sturm–Liouville problem, self-adjointness, eigenfunction expansions/Fourier series etc. are topics far too rich to cover lightly.

But this does not mean the subject should not be mentioned, if for no other reason to point out the big differences between the initial value problem and the boundary value problem. The big distinction is of course that we are requiring the solution to satisfy end point conditions at two distinct points as opposed to one initial point. A simple example points this out:

Given the equation $y'' + y = 0$, where $y = y(x)$, how many solutions does it have for the following boundary conditions?

(i) $y(0) = 1, y\left(\frac{\pi}{2}\right) = 1$: since $y(x) = A \sin x + B \cos x$ is the general solution, then we must solve

$$\left. \begin{array}{l} y(0) = A\sin(0) + B\cos(0) = 1 \\ y\left(\dfrac{\pi}{2}\right) = A\sin\left(\dfrac{\pi}{2}\right) + B\cos(\pi/2) = 1 \end{array} \right\} \Rightarrow A = B = 1,$$

so $y(x) = \sin x + \cos x$ is the unique solution.

(ii) $y(0) = 1, y(\pi) = 1$: we obtain

$$y(0) = A\sin(0) + B\cos(0) = B = 1,$$

$$y(\pi) = A\sin(\pi) + B\cos(\pi) = -B = 1,$$

a contradiction, so no solution exists.

(iii) $y(0) = y(2\pi) = 1$: we obtain

$$\left. \begin{array}{l} A\sin(0) + B\cos(0) = 1 \\ A\sin(2\pi) + B\cos(2\pi) = 1 \end{array} \right\} \Rightarrow B = 1, A \text{ arbitrary}$$

so we have an infinite number of solutions $y(x) = A\sin x + \cos x$.

Note: for boundary value problems, the independent variable is usually x (space) as opposed to t (time).

Another problem worth mentioning is the eigenvalue problem; an example is this:
Given the boundary value problem

$$y'' + \lambda y = 0, \quad y(0) = y(\pi) = 0,$$

for what values of the parameter λ will it have nontrivial solutions? The example indicates that the question of existence and uniqueness of solutions of a boundary value problem is a thorny one.

i) $\lambda < 0$: Letting $\lambda = -r^2, r > 0$, the general solution is

$$y(x) = Ae^{-rx} + Be^{rx}.$$

Then $y(0) = 0$ implies $A + B = 0$, and $y(\pi) = 0$ gives

$$y(\pi) = Ae^{-r\pi} + (-A)e^{r\pi} = 0.$$

If $A = 0$ we have the trivial solution $y(x) \equiv 0$, otherwise $A(e^{-r\pi} - e^{r\pi}) = 0$ gives $1 - e^{2r\pi} = 0$, which is impossible. No solutions.

ii) $\lambda = 0$: then $y'' = 0$ or $y(x) = Ax + B$ and the boundary conditions imply that $A = B = 0$.

iii) $\lambda > 0$: let $\lambda = r^2$, $r > 0$, then $y(x) = A\sin rx + B\cos rx$; $y(0) = 0 \Rightarrow B = 0$, and $y(\pi) = \sin r\pi = 0$ implies $\lambda = 1, 4, \ldots, n^2, \ldots$, the *eigenvalues* of the differential equation with associated *eigenfunctions* $\sin x, \sin 2x, \ldots$, $\sin nx, \ldots$.

This is a good point to stop, otherwise one is willy nilly drawn into the fascinating topic of Fourier series.

Remark. If you are driven to do some numerical analysis, taking advantage of today's accessible computer power, consider doing some numerical approximations of boundary value problems using the shooting method. It has a nice geometric rationale, and one can use bisection or the secant method at the other end to get approximations of the slope. The book by Burden and Faires has a very good discussion. Some sample problems:

i) $y'' + e^{2x}y' + y = 0, y(0) = 0, y(1) = 0.5$

$$\text{(initial guess } y'(0) = 1.2)$$

ii) $y'' + 2y' + xy = \sin x, y(0) = 0, y(1) = 2$

$$\text{(initial guess } y'(0) = 4.0)$$

iii) $y'' - \frac{3}{4}y^5 = 0, y(0) = 2/3, y(7/4) = 1/2$

$$\text{(initial guess } y'(0) = -0.15)$$

Stability

This is a good spot to briefly mention the stability of the solution $x(t) \equiv 0$ of the constant coefficient equation $\ddot{x} + a\dot{x} + bx = 0$. Going back to the mass spring system we see that if $x(t) \equiv \dot{x}(t) \equiv 0$, the system is at rest—this is an equilibrium point. From the nature of solutions we see that

a. If the roots λ of the characteristic polynomial have negative real parts then every solution approaches 0 as $t \to \infty$. The cases are

$$\lambda_1, \lambda_2 < 0: \quad x(t) = c_1 e^{\lambda_1 t} + c_2 e^{\lambda_2 t}$$

$$\lambda < 0 \text{ double root}: \quad x(t) = c_1 e^{\lambda t} + c_2 t e^{\lambda t}$$

$$\lambda = r \pm i\alpha, r < 0: \quad x(t) = A e^{rt} \cos(\alpha t - \varphi).$$

In each case we see that for any $\epsilon > 0$ if $|x(t) - 0|$ is sufficiently small for $t = t_1$ then $|x(t) - 0| < \epsilon$ for all $t > t_1$, and furthermore $\lim_{t \to \infty} |x(t)| = 0$. We conclude that

If the roots of the characteristic polynomial have negative real parts then the solution $x(t) \equiv 0$ is asymptotically stable.

b. If the roots of the roots λ of the characteristic polynomial are purely imaginary $\lambda = \pm i\alpha$, then the solution is $x(t) = A\cos(\alpha t - \varphi)$. We see that if A, which is always positive, satisfies $A < \epsilon$, then $|x(t) - 0| = |A\cos(\alpha t - \varphi)| \le |A| < \epsilon$, so as before, if $|x(t) - 0|$ is sufficiently small for $t = t_1$ then $|x(t) - 0| < \epsilon$ for all $t > t_1$, but $\lim_{t \to \infty} |x(t)| \ne 0$.

We conclude that

If the roots of the characteristic polynomial have zero real parts then the solution $x(t) \equiv 0$ is stable.

From the argument above we see that for a solution to be asymptotically stable it must first be stable—an important point that is sometimes overlooked.

c. Finally in the case where one or both of the roots of the characteristic polynomial have a positive real part we say that the solution $x(t) \equiv 0$ is unstable.

We can use a default definition that not stable means unstable, and clearly in the cases $\lambda_1, \lambda_2 > 0$, double root $\lambda > 0$, or $\lambda = \alpha + i\beta$, $\alpha > 0$, we see that any solution $x(t)$ becomes unbounded as $t \to \infty$ no matter how small its amplitude. What about the case $\lambda_1 < 0 < \lambda_2$? While there is a family of solutions $c_1 e^{\lambda_1 t}$ which approach zero as $t \to \infty$, the remainder are of the form $c_2 e^{\lambda_2 t}$ or $c_1 e^{\lambda_1 t} + c_2 e^{\lambda_2 t}$ which become unbounded. Given any $\epsilon > 0$ we can always find solutions satisfying $|x(0)| < \epsilon$ where $x(0) = c_2$ or $x(0) = c_1 + c_2$ and $c_2 \ne 0$. Such solutions become unbounded as $t \to \infty$ so the homespun stability criterion—once close always close—is not satisfied.

It is important to go through the cases above because they are an important precursor to the discussion of stability of equilibria in the phase plane of a linear or almost linear two-dimensional autonomous systems. In fact, they serve as excellent and easily analyzed models of the various phase plane portraits (node, spiral, saddle, center) of constant coefficient systems. This will be demonstrated in the next chapter.

Infinite Series Solutions

Perhaps the author's opinion about power series solutions is best summarized by a footnote found in Carrier and Pearson's succinct little book on ordinary differential equations, at the start of a short *ten page* chapter on power series. The Chapter 11 referred to is an *eleven page* chapter with some of the important identities and asymptotic and integral representations of the more common special functions—Error, Bessel, Airy, and Legendre.

> *For example, the trigonometric function, $\sin x$, is an elementary function from the reader's point of view because he recalls that $\sin x$ is that odd, oscillatory, smooth function for which $|\sin x| \leq 1$, $\sin x = \sin(x + 2\pi)$ and for which meticulous numerical information can be found in easily accessible tables; the Bessel function, $J_0(x)$, on the other hand, probably won't be an elementary function to that same reader until he has thoroughly digested Chapter 11. Then, $J_0(x)$ will be that familiar, even, oscillatory, smooth function which, when $x \gg 1$, is closely approximated by $\sqrt{2/\pi x} \, \cos(x - \pi/4)$, which obeys the differential equation, $(xu')' + xu = 0$, for which $J_0(0) = 1$, and for which meticulous numerical information can be found in easily accessible tables.

The statement was made in 1968 when practicing mathematicians used *tables* (remember *tables*?), long before the creation of the computing power we enjoy today. So why do we need to have books with 30–50 page chapters of detailed, boring constructions of power series for all the various cases—ordinary point, regular singular point, regular singular point—logarithmic case—followed by another lengthy presentation on the power series representations of the various special functions?[1] Unless you intend to become a specialist in approximation theory or special functions, you are not likely to have to compute a series solution for a nonautonomous second order differential equation in your entire life.

Somebody long ago, maybe Whittaker and Watson, decided that the boot camp of ordinary differential equations was series solutions and special functions, including the Hydra of them all, the hypergeometric function. This makes no sense today and the entire subject should be boiled down to its essence, which is to be familiar with how to construct a series solution, and a *recognition* of those special functions which come up in partial differential equations and applied mathematics via eigenfunction expansions.

The point made in the quote above is an important one, namely that there are differential equations whose *explicit* solutions are special functions, whose various properties such

---

[1] The author labored long and hard, seeking enlightenment, writing such a chapter in a textbook some time ago. Whatever nascent interest he had in the topic was quickly throttled.

as orthogonality and oscillatory behavior are well known, and whose representations are convergent power series with known coefficients. For instance:

The Bessel function of order 2

$$J_2(x) = \left(\frac{x}{2}\right)^2 \sum_0^\infty \frac{(-1)^n}{n!(n+2)!} \left(\frac{x}{2}\right)^n$$

is a solution of the Bessel Equation of order 2,

$$\frac{d^2y}{dx^2} + \frac{1}{x}\frac{dy}{dx} + \frac{x^2 - 2^2}{x^2}y = 0.$$

The function $J_2(x)$ is an oscillatory function with an infinite number of zeros on the positive x-axis, and $J_2(x) \to 0$ as $x \to \infty$. It is well tabulated.

Hence, we can certainly say that we can *solve* the Bessel equation of order 2, although the solution is not in terms of polynomials or elementary functions. Power series solutions do give us a broader perspective on what is the notion of the solution of an ODE.

Introduce the topic with the construction of the first few terms of the series for $Ai(x)$ or $J_0(x)$, then develop the recurrence relation for their terms. For the same reason you should not use Euler's Method to approximate the solutions of $\dot{x} = x$, do not develop the series expansion for the solutions of $\ddot{x} + x = 0$; it gives the subject a bad reputation.

Next, if you wish, discuss briefly some of the special functions, their differential equations, and possibly some of the orthogonality relations, but this is already wandering far afield. Then move on, remembering that if someone wants the graph of $J_{7/2}(x)$ or its value at $x = 11.054$ a computer package or even some hand calculators will provide it—but who would ask such a question?

Something worth considering instead of power series expansions is *perturbation expansions*; these are of great importance in applied mathematics and in the analysis of periodic solutions of nonlinear ODEs. See for instance, the Poincaré–Lindstedt method in more advanced texts. First order equations can provide some nice examples as the following shows:

Consider the logistic equation with periodic harvesting

$$\dot{x} = \frac{1}{10}x(1 - x/40) - \left(\frac{3}{4} + \epsilon \sin t\right).$$

Then $r = \frac{1}{10}, K = 40$ so $rK/4 = 1$, the critical harvesting level, so for existence of periodic solutions we require that $\left|\frac{3}{4} + \epsilon \sin t\right| < 1$ or $|\epsilon| < \frac{1}{4}$. We wish to approximate the stable periodic solution with a power series in ϵ having periodic coefficients, so let

$$x(t) = x_0(t) + \epsilon x_1(t) + \epsilon^2 x_2(t) + \cdots, x_i(t) = x_i(t + 2\pi).$$

Substitute and equate like powers of ϵ:

ϵ^0: $\dot{x}_0 = \frac{1}{10}x_0 - \frac{1}{400}x_0^2 - \frac{3}{4}$. The only periodic solutions will be constant ones $x_0(t) \equiv 10, 30$. These correspond to the new equilibria under constant rate harvesting $H = \frac{3}{4}$. To perturb around the stable one let $x_0(t) \equiv 30$.

$\epsilon^1:\ \dot{x}_1 = \left(\frac{1}{10} - \frac{1}{200}x_0\right)x_1 - \sin t = -\frac{1}{20}x_1 - \sin t$. The solution is

$$x_1(t) = A_1 e^{-t/20} + \frac{20}{401}(20\cos t - \sin t)$$

$$= A_1 e^{-t/20} + \frac{20}{\sqrt{401}}\cos(t + \phi),\ \phi = \tan^{-1}(1/20).$$

Since we want periodicity let $A_1 = 0$, and note that the exponentially decaying term supports the fact that $x(t)$ is stable.

$\epsilon^2:\ \dot{x}_2 = \left(-\frac{1}{10} - \frac{1}{200}x_0\right)x_2 - \frac{1}{400}x_1^2 = -\frac{1}{20}x_2 - \frac{1}{401}\cos^2(t + \phi)$, which has a periodic solution (left to the reader).

We conclude that

$$x(t) \approx 30 + \frac{20\epsilon}{\sqrt{401}}\cos(t + \phi) + O(\epsilon^2).$$

These kinds of expansions are very much in the spirit of the qualitative theory of ODEs.

5

Linear and Nonlinear Systems

This final chapter will not be a lengthy one, despite the importance of the topic. The reason is a refreshing one: many of today's textbooks are bulging with nonlinear systems and the models describing them. This is largely the result of the great interest today in dynamical systems, and the availability of the computing power to be able to display and approximate solutions. While the author may question some of the topics chosen or the level of sophistication, it is all done in the modern spirit of qualitative understanding, and that merits approbation.

1 Constant Coefficient Linear Systems

In the previous chapter on second order equations, the linear systems approach, it was shown that for the two-dimensional system

$$\dot{x} = a(t)x + b(t)y, \quad \dot{y} = c(t)x + d(t)y$$

the solution is given by $\underset{\sim}{x}(t) = \Phi(t)\underset{\sim}{c}$ where

$$\underset{\sim}{x}(t) = \begin{pmatrix} x(t) \\ y(t) \end{pmatrix}, \quad \Phi(t) = \begin{pmatrix} x_1(t) & x_2(t) \\ y_1(t) & y_2(t) \end{pmatrix}$$

with $\mathrm{col}(x_1(t), y_1(t))$, $\mathrm{col}(x_2(t), y_2(t))$ a fundamental pair of solutions, and $\underset{\sim}{c}$ a constant vector determined by the initial conditions, if any. This result generalizes to higher dimensions; it is largely of theoretical interest since finding fundamental sets of solutions will be nearly impossible to accomplish.

For the constant coefficient case

$$\dot{\underset{\sim}{x}} = A\underset{\sim}{x}, \quad A = (a_{ij}), \quad i, j = 1, \dots, n,$$

one needs to know that if λ is an eigenvalue of A then $\underset{\sim}{x} = e^{\lambda t}\underset{\sim}{\eta}$ is a solution, where $\underset{\sim}{\eta}$ is an eigenvector corresponding to λ. A one line proof suffices:

$$\underset{\sim}{\dot{x}} = \lambda e^{\lambda t}\underset{\sim}{\eta} = A e^{\lambda t}\underset{\sim}{\eta} \Leftrightarrow e^{\lambda t}(A - I\lambda)\underset{\sim}{\eta} = 0.$$

But where to go from here to get $\Phi(t)$? There are lots of perilous paths, many requiring a real knowledge of linear algebra (Jordan canonical form, companion matrices, etc.), and the reason is simply because A might have eigenvalues of multiplicity larger than one.

Spending a lot of time on what is basically a linear algebra problem seems a waste of time since

(i) We know a fundamental set of solutions exists—let $x_i(t)$ be the unique solutions of the IVP $\underset{\sim}{\dot{x}} = A\underset{\sim}{x}$, $\underset{\sim}{x}(0) = \underset{\sim}{e}_i$, the ith unit vector, etc.

(ii) We can do a little hand waving and state the plausible result

If the $n \times n$ matrix A has n distinct eigenvalues λ_i with associated eigenvectors η_i, $i = 1, 2, \ldots, n$ then the set $\{\underset{\sim}{x}_i(t)\} = \{e^{\lambda_i t}\underset{\sim}{\eta}_i\}$ is a fundamental set of solutions of the system $\underset{\sim}{\dot{x}} = A\underset{\sim}{x}$.

(iii) We can quickly take care of the case $n = 2$ completely, and it gives a hint of what the problem will be for higher dimensions when multiplicities occur.

(iv) The subject is ordinary differential equations not linear algebra.

Recently, some textbooks have appeared doing a lot of linear algebra for the case $n = 3$ and then discussing three dimensional systems. The impetus for this comes from the availability of better 3-D graphics, and because of the great interest in the three dimensional systems exhibiting chaotic behavior—the Lorenz equations modeling climate change, for instance. Unfortunately, two-dimensional systems cannot have chaotic structures, but that isn't a good reason to spend excessive time analyzing 3×3 matrices.

Returning to the cozy $n = 2$ world: consider

$$\underset{\sim}{\dot{x}} = \begin{pmatrix} \dot{x} \\ \dot{y} \end{pmatrix} = \begin{pmatrix} a & b \\ c & d \end{pmatrix} \begin{pmatrix} x \\ y \end{pmatrix} = A\underset{\sim}{x}$$

and we can state:

If λ_1, λ_2 are distinct eigenvalues with associated eigenvectors η_1, η_2, then $\underset{\sim}{x}_1(t) = e^{\lambda_1 t}\underset{\sim}{\eta}_1$, $\underset{\sim}{x}_2(t) = e^{\lambda_2 t}\underset{\sim}{\eta}_2$ are a fundamental pair of solutions and every solution can be written as

$$\underset{\sim}{x}(t) = c_1\underset{\sim}{x}_1(t) + c_2\underset{\sim}{x}_2(t)$$

for appropriate choices of the constants c_1, c_2.

What about the case of complex conjugate eigenvalues $\lambda = r + iq$, $\bar{\lambda} = r - iq$, $q \neq 0$? For any real matrix A, if λ is a complex eigenvalue with associated complex eigenvector η then since $\bar{A} = A$,

$$(A - \lambda I)\eta = 0 \Leftrightarrow (A - \bar{\lambda}I)\bar{\eta} = 0,$$

so $\bar{\eta}$ is an eigenvector associated with $\bar{\lambda}$, and the statement above applies.

But we want real solutions, so go back to a variation of the technique used for the second order equation whose characteristic polynomial had complex roots. If $\lambda = r + iq$ and $\underset{\sim}{\eta} = (\alpha + i\beta, \gamma + i\delta)$ then any solution can be expressed as

$$\underset{\sim}{x}(t) = e^{rt} \left[c_1 e^{iqt} \begin{pmatrix} \alpha + i\beta \\ \gamma + i\delta \end{pmatrix} + c_2 e^{-iqt} \begin{pmatrix} \alpha - i\beta \\ \gamma - i\delta \end{pmatrix} \right].$$

Now let $c_1 = c_2 = \frac{1}{2}$, then $c_1 = \frac{1}{2i}$, $c_2 = -\frac{1}{2i}$, and express $\cos qt$, $\sin qt$ in terms of complex exponentials to get two real solutions

$$\underset{\sim}{x}_1(t) = e^{rt} \begin{pmatrix} \alpha \cos qt - \beta \sin qt \\ \gamma \cos qt - \delta \sin qt \end{pmatrix}$$

$$\underset{\sim}{x}_2(t) = e^{rt} \begin{pmatrix} \alpha \sin qt + \beta \cos qt \\ \gamma \sin qt + \delta \sin qt \end{pmatrix},$$

and there's our fundamental pair.

The case of λ a double root is vexing when we can only find one eigenvector $\underset{\sim}{\eta}$. The linear algebra approach is to find a second nonzero vector $\underset{\sim}{\sigma}$ satisfying $(A - \lambda I)\underset{\sim}{\sigma} = \underset{\sim}{\eta}$. Then a fundamental pair of solutions is

$$\underset{\sim}{x}_1(t) = e^{\lambda t}\underset{\sim}{\eta}, \quad \underset{\sim}{x}_2(t) = e^{\lambda t}(\underset{\sim}{\sigma} + \underset{\sim}{\eta}t),$$

and this must be proved by showing $\underset{\sim}{\sigma}$ is not a multiple of $\underset{\sim}{\eta}$. We will do this shortly.

But a differential equations approach is more intuitive, suggested by the strategy used in the second order case of multiplying by t. Try a solution of the form $te^{\lambda t}\underset{\sim}{\eta}$—it doesn't work. So try instead that solution plus $\underset{\sim}{\sigma}e^{\lambda t}$ where $\underset{\sim}{\sigma}$ is a nonzero vector. Then $\underset{\sim}{x}(t) = te^{\lambda t}\underset{\sim}{\eta} + e^{\lambda t}\underset{\sim}{\sigma}$ is assumed to be a solution, and since $\dot{\underset{\sim}{x}} = A\underset{\sim}{x}$, a little calculation gives

$$\dot{\underset{\sim}{x}} = \lambda te^{\lambda t}\underset{\sim}{\eta} + e^{\lambda t}(\underset{\sim}{\eta} + \lambda\underset{\sim}{\sigma}) = te^{\lambda t}A\underset{\sim}{\eta} + e^{\lambda t}A\underset{\sim}{\sigma}.$$

But $A\underset{\sim}{\eta} = \lambda\underset{\sim}{\eta}$ so the first terms cancel and after cancelling $e^{\lambda t}$ we are left with

$$A\underset{\sim}{\sigma} = \underset{\sim}{\eta} + \lambda\underset{\sim}{\sigma} \quad \text{or} \quad (A - \lambda I)\underset{\sim}{\sigma} = \underset{\sim}{\eta}$$

which is exactly what we want.

If $\underset{\sim}{\sigma}$ were a multiple of $\underset{\sim}{\eta}$, say $\underset{\sim}{\sigma} = k\underset{\sim}{\eta}$, then the equations $(A - \lambda I)k\underset{\sim}{\eta} = \underset{\sim}{\eta}$ would imply $k(A - \lambda I)\underset{\sim}{\eta} = 0 = \underset{\sim}{\eta}$ which is impossible. Now we can prove $\underset{\sim}{x}_1(t)$ and $\underset{\sim}{x}_2(t)$ are a fundamental pair of solutions by showing we can use them to solve uniquely any IVP, $\dot{\underset{\sim}{x}} = A\underset{\sim}{x}$, $\underset{\sim}{x}(0) = \text{col}(x_0, y_0)$ with a linear combination

$$\underset{\sim}{x}(t) = c_1\underset{\sim}{x}_1(t) + c_2\underset{\sim}{x}_2(t) = c_1 e^{\lambda t}\underset{\sim}{\eta} + c_2 e^{\lambda t}(\underset{\sim}{\sigma} + t\underset{\sim}{\eta}).$$

If $\underset{\sim}{\eta} = \text{col}(u, v)$, $\underset{\sim}{\sigma} = \text{col}(w, z)$ this leads to the equations

$$c_1 u + c_2 w = x_0, \quad c_2 v + c_2 z = y_0$$

which can be solved for any x_0, y_0 since the columns of the coefficient matrix $\left(\begin{smallmatrix} u & w \\ v & z \end{smallmatrix} \right)$ are linearly independent, so its determinant is not zero.

All this takes longer to write than it does to teach, but limiting the discussion to $n = 2$ makes the calculations easy and avoids all the problems associated with dimensionality. Just consider $n = 3$ with λ an eigenvalue of multiplicity three and the various possibilities; you might shudder and just leave it to the linear algebra computer package.

For the nonhomogeneous system

$$\dot{\underset{\sim}{x}} = A(t)\underset{\sim}{x} + \underset{\sim}{B}(t) \quad \text{or} \quad \dot{\underset{\sim}{x}} = A\underset{\sim}{x} + \underset{\sim}{B}(t)$$

we have discussed the general variation of parameters formula for $n = 2$ in the previous chapter. It generalizes easily to higher dimensions. In the case $A(t) = A$ note that the variation of parameters formula does not require the computation of $\Phi^{-1}(s)$ when $\Phi(s)$ is the fundamental matrix satisfying $\Phi(0) = I$. This is because $\Phi(t)\Phi^{-1}(s) = \Phi(t - s)$ which is not hard to prove:

Let $\Omega(t) = \Phi(t)\Phi^{-1}(s)$ and since $\dot{\Phi} = A\Phi$ then $\dot{\Omega}(t) = \dot{\Phi}(t)\Phi^{-1}(s) = A\Phi(t)\Phi^{-1}(s) = A\Omega(t)$. Therefore $\Omega(t)$ is the solution of the IVP $\dot{X} = AX$, $X(s) = I$, but so is $\Phi(t - s)$ hence by uniqueness $\Omega(t) = \Phi(t - s)$.

Finally we note that the method of comparison of coefficients can be used to solve $\dot{\underset{\sim}{x}} = A\underset{\sim}{x} + \underset{\sim}{B}(t)$ if $\underset{\sim}{B}(t)$ has the appropriate form, but it can be a bookkeeping nightmare. For example to find a particular solution for the system

$$\begin{pmatrix} \dot{x} \\ \dot{y} \end{pmatrix} = A \begin{pmatrix} x \\ y \end{pmatrix} + \begin{pmatrix} t^2 \\ \sin t \end{pmatrix} \quad \text{try} \quad \begin{cases} x_p(t) = Bt^2 + Ct + D + E\sin t + F\cos t \\ y_p(t) = Gt^2 + Ht + J + K\sin t + L\cos t \end{cases}$$

substitute and compare coefficients, and that is assuming $\sin t$ is not one of the components of the solutions of $\dot{\underset{\sim}{x}} = A\underset{\sim}{x}$. Good luck—if you do the problem three times and get the same answer twice—it's correct.

2   The Phase Plane

The phase plane is the most valuable tool available to study *autonomous* two-dimensional systems

$$(\star) \qquad\qquad \dot{x} = P(x, y) \quad \dot{y} = Q(x, y),$$

and it is important to stress that the right-hand sides depend only on x and y—there is no t-dependence. If there is t-dependence so that $P = P(x, y, t)$, $Q = Q(x, y, t)$ the system can be transformed to a 3-dimensional autonomous system by introducing a third variable $z = t$. Then we have

$$\dot{x} = P(x, y, z), \quad \dot{y} = Q(x, y, z), \quad \dot{z} = 1,$$

which sometimes can be useful and introduces 3D-graphs of solutions. We will pass on this.

For simplicity, we will assume that $P(x, y), Q(x, y)$ are continuous together with their first partial derivatives for all x, y. The very important point is that given a solution

$(x(t), y(t))$ of $(\star)$, then as t varies it describes parametrically a curve in the x, y-plane. This curve is called a *trajectory or orbit* for the following simple reason which is an easy consequence of the chain rule:

> If $(x(t), y(t))$ is a solution of $(\star)$ and c is any real constant then $(x(t+c), y(t+c))$ is also a solution.

But if $(x(t), y(t))$ describes a trajectory then $(x(t + c), y(t + c))$ describes the same parametric curve, but with the time shifted, and hence describes the same trajectory. Quick, an example, Herr Professor!

> $\dot{x} = y, \dot{y} = -4x$ (corresponding to $\ddot{x} + 4x = 0$). If $x(0) = 1$, $y(0) = 0$ then $x(t) = \cos 2t$, $y(t) = -2 \sin 2t$ so $x(t)^2 + \frac{y(t)^2}{4} = 1$. The trajectory is an ellipse traveled counterclockwise and we have the phase plane picture

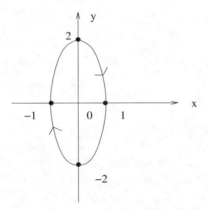

This trajectory represents every solution that satisfies the initial conditions $x(t_0) = x_0$, $y(t_0) = y_0$, where $x_0^2 + \frac{y_0^2}{4} = 1$. For instance, $x(t) = \cos 2(t - \pi/2)$, $y(t) = -2 \sin 2(t - \pi/2)$ lies on the ellipse and satisfies the initial conditions $x\left(\frac{\pi}{2}\right) = 1$, $y(\pi/2) = 0$, or the initial conditions $x\left(\frac{\pi}{3}\right) = \frac{1}{2}$, $y\left(\frac{\pi}{3}\right) = \sqrt{3}$.

Remark. To determine the direction of travel it is usually easy to let $x > 0$ and $y > 0$ and determine the direction of growth. In the example above if $y > 0$ then $\dot{x} > 0$ so the x-coordinate increases, and if $x > 0$ then $\dot{y} < 0$ so the y-coordinate decreases.

The next important point is that trajectories don't cross. If they did you could reparametrize by shifting time, and create the situation where two distinct solutions passed through the same point at the same time. This would violate uniqueness of solutions of the IVP.

Our phase plane is now filled with nonintersecting trajectories waltzing across it— "filled" is right since for every point (x_0, y_0) and any time t_0 we can find a solution of $(\star)$ with $x(t_0) = x_0$, $y(t_0) = x_0$. What else could it contain? Points (x_0, y_0) where $P(x_0, y_0) = Q(x_0, y_0) = 0$ called *equilibrium points*, or *critical points*. If we think of the system $(\star)$ as describing motion then an equilibrium point would be a point where $\dot{x} = \dot{y} = 0$ so the system is at rest.

What else? It could contain closed curves whose trajectories represent periodic solutions. For such a solution it must be the case that $(x(t+T), y(t+T)) = (x(t), y(t))$ for all t, and some minimal $T > 0$ called its period. Note that since the equations are autonomous, the period T is unspecified so it must be calculated or estimated from the differential equation. This is often a formidable task when P and Q are nonlinear.

Closed trajectories are called *cycles* and a major part of the research done in differential equations in the last century was the search for *isolated* ones or *limit cycles*. By isolated is meant there is some narrow band enclosing the closed trajectory which contains no other closed trajectory. An informal definition would be that for a limit cycle, nearby solutions could spiral towards it (a stable limit cycle), or spiral away from it (an unstable limit cycle), or both (a semistable limit cycle).

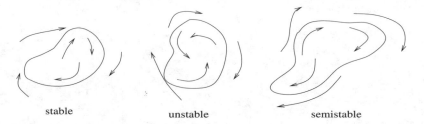

stable   unstable   semistable

The important point to be made about the phase plane is that the nonintersecting trajectories, possibly equilibria, cycles, and possibly limit cycles, are all its possible inhabitants. This is the result of some fairly deep analysis and that the universe being studied is the xy-plane. It has the special property that a closed, simple curve (it doesn't overlap itself) divides the plane into an "outside" and "inside", and this restricts the kind of behavior we can expect from solutions of $(\star)$. Nevertheless, the phase plane can be spectacular in its simplicity.

3   The Linear System and the Phase Plane

For the 2-dimensional linear system

$$\dot{x} = ax + by, \quad \dot{y} = cx + dy,$$

clearly (0,0) is an equilibrium point and to ensure it is the only one we stipulate that $ad - bc \neq 0$. The case where $ad - bc = 0$ will be mentioned briefly later. A discussion of this usually follows some analysis of linear systems, so expositors are eager to use all the stuff they know about $\dot{\underset{\sim}{x}} = A\underset{\sim}{x}$ where $A = \left(\begin{smallmatrix} a & b \\ c & d \end{smallmatrix}\right)$.

This often confuses matters and it is much easier to use the second order equation,

$$\ddot{x} + p\dot{x} + qx = 0, \quad A = \begin{pmatrix} 0 & 1 \\ -q & -p \end{pmatrix},$$

since we know all about its solutions and we can describe every possible phase plane configuration for the general linear system, except one. We analyze the various cases of roots λ_1, λ_2 of the characteristic polynomial $p(\lambda) = \lambda^2 + p\lambda + q$, and the *important point* is that the pictures for the full linear system will be qualitatively the same, modulo a rotation or dilation.

1. $-\lambda_1, -\lambda_2$ negative and distinct.

$$x(t) = c_1 e^{-\lambda_1 t} + c_2 e^{-\lambda_2 t}$$
$$\dot{x}(t) = y(t) = -\lambda_1 c_1 e^{-\lambda_1 t} - \lambda_2 c_2 e^{-\lambda_2 t}$$

$c_2 = 0$: $(x(t), y(t)) \to (0,0)$ as $t \to \infty$, and $y(t) = -\lambda_1 x(t)$; the trajectory is a straight line going into the origin.
$c_1 = 0$: $(x(t), y(t)) \to (0,0)$ as $t \to \infty$, and $y(t) = -\lambda_2 x(t)$; the trajectory is a straight line going into the origin.
Now suppose $\lambda_2 > \lambda_1$ then

$$\frac{y(t)}{x(t)} = \frac{-\lambda_1 c_1 - \lambda_2 c_2 e^{-(\lambda_2 - \lambda_1)t}}{c_1 + c_2 e^{-(\lambda_2 - \lambda_1)t}} \to -\lambda_1 \text{ as } t \to \infty$$

so all trajectories are asymptotic to the line $y = -\lambda_1 x$ and we get this picture—the origin is a *stable node* which is asymptotically stable.

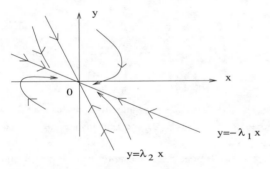

Note that the straight line trajectories do intersect at the origin when $t = \infty$ which does not contradict the statement that trajectories do not intersect.

2. λ_1, λ_2 distinct and positive. The previous picture is the same but the arrows are reversed—the origin is an *unstable node*.

3. λ_1, λ_2 complex conjugate with negative real part, so $\lambda_{1,2} = r \pm i\alpha, r < 0$. Use the phase-amplitude form of the solution:

$$x(t) = A e^{-rt} \cos(\alpha t - \phi),$$
$$\dot{x}(t) = y(t) = -r A e^{-rt} \cos(\alpha t - \phi) - \alpha A e^{-rt} \sin(\alpha t - \phi),$$

so clearly $(x(t), y(t)) \to (0,0)$ as $t \to \infty$. Now do a little manipulation:

$$\frac{y(t) + rx(t)}{\alpha} = -A e^{-rt} \sin(\alpha t - \phi)$$

and therefore the trajectory is represented by the equation $\left(\frac{y+rx}{\alpha}\right)^2 + x^2 = A^2 e^{-2rt}$,

and expanding

$$y^2 + 2rxy + (\alpha^2 + r^2)x^2 = \alpha^2 A^2 e^{-2rt}, \quad r > 0.$$

In times past, when analytic geometry was not a brief appendix in a calculus text, one would know that a rotation of axes would convert the left-hand side to an elliptical form. Since the right-hand side is a positive "radius" approaching zero as $t \to \infty$—the origin is a *stable spiral* which is asymptotically stable.

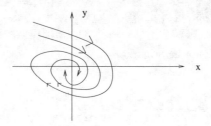

4. λ_1, λ_2 complex conjugate with positive real part. The previous picture is the same but the arrows are reversed—the origin is an *unstable spiral*.

5. $-\lambda_1 < 0 < \lambda_2$.

$$x(t) = c_1 e^{-\lambda_1 t} + c_2 e^{\lambda_2 t}$$
$$\dot{x}(t) = y(t) = -\lambda_1 c_1 e^{-\lambda_1 t} + \lambda_2 c_2 e^{\lambda_2 t}$$

$c_2 = 0$: $(x(t), y(t)) \to (0,0)$ as $t \to \infty$ and $y(t) = -\lambda_1 x(t)$
$c_1 = 0$: $(x(t), y(t)) \to (0,0)$ as $t \to -\infty$ and $y(t) = \lambda_2 x(t)$

We get two straight line trajectories, one entering the origin and the other leaving and the remaining trajectories approach them as $t \to \pm\infty$—the origin is an *unstable saddle point*.

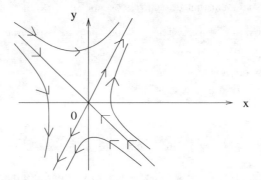

6. We have previously discussed the case $\lambda = \pm i\alpha$.

$$\left.\begin{array}{l} x(t) = A\cos(\alpha t - \phi) \\ \dot{x}(t) = y(t) = -\alpha A \sin(\alpha t - \varphi) \end{array}\right\} \quad x^2 + \frac{y^2}{\alpha^2} = A^2$$

The trajectories are families of ellipses—the origin is a *center* which is stable, but not asymptotically stable.

Note that the trajectories are cycles since they are closed curves, but not isolated ones so they are not limit cycles.

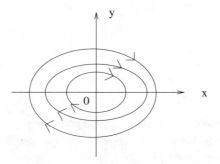

7. $-\lambda < 0$, a double root.

$$x(t) = c_1 e^{-\lambda t} + c_2 t e^{-\lambda t}$$

$$\dot{x}(t) = y(t) = -\lambda c_1 e^{-\lambda t} - \lambda c_2 t e^{-\lambda t} + c_2 e^{-\lambda t} = -\lambda x(t) + c_2 e^{-\lambda t}$$

Then $(x(t), y(t)) \to (0,0)$ as $t \to \infty$, and if $c_2 = 0$ the trajectory is a straight line entering the origin. If $c_2 \neq 0$ all trajectories are asymptotic to that line—the origin is a *stable node* which is asymptotically stable.

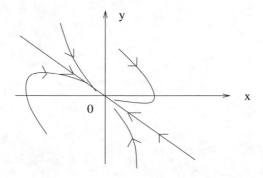

8. $\lambda > 0$, a double root. The previous picture is the same but the arrows are reversed— the origin is an unstable node.

There is a degenerate case corresponding to the second order equation $\ddot{x} + a\dot{x} = 0$, so the eigenvalues are $\lambda = 0$, $\lambda = -a$. Then

$$x(t) = c_1 + c_2 e^{-at}, \quad \dot{x}(t) = y(t) = -ac_2 e^{-at}$$

and suppose $a > 0$. The $(x(t), y(t)) \to (c_1, 0)$ as $t \to \infty$ and $x(t) = c_1 - \frac{1}{a}y(t)$ or $y(t) = -ax(t) + ac_1$ a family of straight lines. The picture is

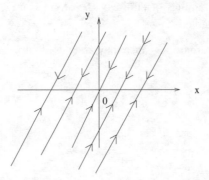

and it implies every point on the x-axis is a critical point. If $a < 0$ the arrows are reversed. There are several other degenerate cases corresponding to the case where $\det A = 0$, but the condition $\det A \neq 0$ assures us that $(0,0)$ is the only equilibrium point.

Using the second order equation to model the various cases does not pick up one case found in the full linear system. It corresponds to the system

$$\dot{x} = qx, \quad \dot{y} = qy \Rightarrow x(t) = c_1 e^{qt}, \quad y(t) = c_2 e^{qt}$$

so $y(t) = \frac{c_2}{c_1} x(t)$, and if $q < 0$, then $(x(t), y(t)) \to (0,0)$. The trajectories are straight lines of arbitrary slope going into the origin; if $q > 0$ they leave the origin. These are also called *nodes,* or more specifically, *proper nodes*, and the origin is asymptotically stable if $q < 0$ and unstable if $q > 0$.

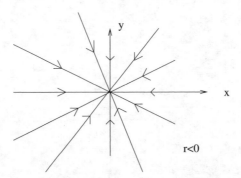

r<0

You will find that the effort to characterize the various cases in the general case is not much more difficult, except that the linear algebra gets in the way. The characteristic polynomial of A is $\lambda^2 - (a + d)\lambda + (ad - bc)$ which is more complicated, and one must find eigenvectors or do a rotation of axes to pull out the pictures.

When one considers the almost linear system or the linear approximation at an equilibrium point of the full system, one should think of the nonlinearity as blurring the pictures above, or tweaking the eigenvalues of the matrix. Hence, one can rationalize that almost always the asymptotically stable or unstable structure will be preserved. It is the stable case, the center, which can be preserved or transformed into a stable or unstable spiral under small perturbation. The zero real part of the eigenvalue can stay the same or shift a little in either direction. Developing this kind of thinking, not rote memorization of formulas or procedures, is the goal.

Afterthought: Just as a topic of academic interest it is very difficult to come up with an example where solutions satisfy $\lim_{t \to \infty} (x(t), y(t)) = (x_0, y_0)$ but (x_0, y_0) is not stable. Solutions must have the property that they get close to (x_0, y_0) then veer away, then get closer, then veer away again etc. so that the "once close, always close" property doesn't hold. One example in polar coordinates $x = r \cos \theta$, $y = r \sin \theta$ is

$$\dot{r} = r(1 - r), \quad \dot{\theta} = \sin^2 \theta,$$

and since $r = 1$ is an attractor, we see that $(r, \theta) = (1, 0)$ is an attractor and $\lim_{t \to \infty} r(t) = 1$. But (1,0) is unstable in view of the behavior of $\theta(t)$ which was mentioned in Section 2.3, and solutions keep leaving any neighborhood of (1,0) temporarily.

4 Competition and Predator-Prey Systems

This section is not intended to give a lengthy description of models of rabbits slugging it out with field mice or trying to escape from coyotes. Rather we will give a rationale for the construction of such models, and mention the interesting case where one of the populations is being harvested.

Given X, Y two populations, with $x(t)$, $y(t)$ their sizes at time t, let $\dot{x}/x$, $\dot{y}/y$ be their *per capita* rate of growth, and assume these are functions of their present size. This leads to the model governing their growth

$$\dot{x} = xf(x, y), \quad \dot{y} = yg(x, y),$$

and we can assume that $f(x, y)$ and $g(x, y)$ are smooth functions for $x \geq 0$, $y \geq 0$ (population sizes aren't negative!). This simple assumption makes it much easier to classify models, and do some preliminary analysis before specifying the nature of f and g.

Now we can distinguish between two models:

Competition Models:

> If y increases then the per capita rate of growth of X decreases. Similarly, if x increases then the per capita rate of growth of Y decreases.

Predator-Prey Models:

> Let Y be a predator population and X a prey population. If x increases then the per capita rate of growth of Y increases, whereas if y increases the per capita rate of growth of X decreases.

Given these assumptions, we will study the stability of any equilibria by examining the associated linearized system at an equilibrium point. Recall that for a nonlinear system

$$\dot{x} = P(x, y), \quad \dot{y} = Q(x, y),$$

and equilibrium point (x_∞, y_∞), where $P(x_\infty, y_\infty) = Q(x_\infty, y_\infty) = 0$, the associated linearized system is $\dot{\underset{\sim}{u}} = A\underset{\sim}{u}$ where $\underset{\sim}{u} = \text{col}(u, v)$,

$$A = \begin{pmatrix} P_x(x,y) & Q_x(x,y) \\ P_y(x,y) & Q_y(x,y) \end{pmatrix}_{(x_\infty, y_\infty)},$$

and $u = x - x_\infty, v = y - y_\infty$.

Competition Models

In this case $f(x, y)$ should decrease as y increases, and $g(x, y)$ should decrease as x increases, and therefore

$$\frac{\partial f}{\partial y}(x, y) < 0, \quad \frac{\partial g}{\partial x}(x, y) < 0, \quad x > 0, y > 0.$$

Furthermore, $f(x, 0)$ and $g(0, y)$ should represent some suitable growth law for each population in the absence of the other.

Types of Equilibria.

The matrix of the linearized system is

$$A = \begin{pmatrix} f(x,y) + xf_x(x,y) & \overset{(<0)}{xf_y(x,y)} \\ \underset{(<0)}{yg_x(x,y)} & g(x,y) + yg_y(x,y) \end{pmatrix}_{(x_\infty, y_\infty)},$$

and the possible equilibria are:

$(\mathbf{0,0})$: We have $\dot{u} = uf(0,0)$, $\dot{v} = vg(0,0)$ and

$$A = \begin{pmatrix} f(0,0) & 0 \\ 0 & g(0,0) \end{pmatrix}.$$

Our basic assumption tells us that u should increase if $v = 0$, and similarly for v, so we must have

$$f(0,0) > 0, \quad g(0,0) > 0,$$

and therefore (0,0) is an unstable node.

The growth of X in the absence of Y ($y = 0$) is governed by the equation $\dot{x} = xf(x, 0)$, so there could be the possibility of a stable equilibrium K (carrying capacity). This would imply that $f(K, 0) = 0$ and $Kf'(K, 0) = Kf_x(K, 0) < 0$. Note: there doesn't have to be such a K e.g., $\dot{x} = ax - bxy$, a, b positive and if $y = 0$ then $\dot{x} = ax$ which is exponential growth.

$(\mathbf{K,0})$: Then

$$A = \begin{pmatrix} 0 + \overset{(<0)}{Kf_x(K,0)} & Kf_y(K,0) \\ 0 & g(K,0) \end{pmatrix}.$$

It could be that $g(K, 0) = 0$ in which case $(K, 0)$ would be a degenerate stable node, but we will ignore that possibility. Then we expect that if $(K, 0)$ were stable then competition would favor X whereas Y would perish, so

$$g(K,0) < 0 \Rightarrow (K,0) \text{ a stable node}.$$

But if $(K,0)$ were unstable then for x near K and y small, population Y would increase, so

$$g(K,0) > 0 \Rightarrow (K,0) \text{ a saddle},$$

$(\mathbf{0}, \mathbf{J})$ **:** Here J plays a similar role as K for X, and the analysis is the same with $g(0,J)$ replacing $f(K,0)$.

We have made considerable inroads on what the phase plane portrait will look like with minimal assumptions on the nature of $f(x,y)$ and $g(x,y)$. At this point the various possibilities are:

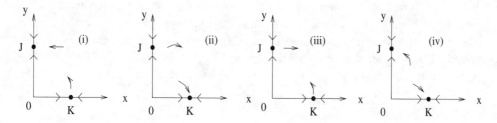

Now, the question becomes whether there exists one or more equilibrium points (x_∞, y_∞), $x_\infty > 0$, $y_\infty > 0$, where $f(x_\infty, y_\infty) = g(x_\infty, y_\infty) = 0$. If such a point were stable we have the case of *coexistence*; if it were unstable we have the case of *competitive exclusion*. In a very complex model there could be both (this might also occur in a simple model if some exterior factor like harvesting were imposed), so that for instance when X is small enough both survive, but if it is too big it wins.

The case where there is one equilibrium (x_∞, y_∞) can now be analyzed given specific forms for $f(x,y)$ and $g(x,y)$. The usual textbook models assume that the growth of each population in the absence of the other is logistic:

$$\dot{x} = rx\left(1 - \frac{x}{K}\right) - x\alpha(x,y), \quad \dot{y} = sy\left(1 - \frac{y}{J}\right) - y\beta(x,y)$$

where r, s, K, and J are all positive, and $\alpha(x,y)$, $\beta(x,y)$ are positive functions for $x > 0$, $y > 0$.

We will stop here with the competition case, but note that the previous graphs already say a lot. In cases (i) and (ii) there need not be a point (x_∞, y_∞) so in (i) Y wins no matter what, and in (ii) X wins. For graph (iii) we suspect that there would be a stable equilibrium (x_∞, y_∞), whereas in (iv) it would be an unstable one.

Predator-Prey Models

Recall that the model is

$$\dot{x} = xf(x,y), \quad \dot{y} = yg(x,y)$$

and the assumption that the per capita rate of growth of the prey X decreases as the predator Y increases, whereas the per capita growth of Y increases as X increases.

Therefore

$$\frac{\partial f}{\partial y}(x,y) < 0, \quad \frac{\partial g}{\partial x}(x,y) > 0, \qquad x > 0, y > 0$$

and the matrix A is

$$A = \begin{pmatrix} f(x,y) + xf_x(x,y) & \overset{(<0)}{xf_y(x,y)} \\ \underset{(>0)}{yg_x(x,y)} & g(x,y) + yg_y(x,y) \end{pmatrix}_{(x_\infty, y_\infty)}.$$

Further natural assumptions would be that X increases if $y = 0$, hence the prey grows in the absence of predators, and if X is the sole or major food supply of Y, then Y goes extinct when $x = 0$.

These assumptions mean that the equilibrium point $(0,0)$ must be a saddle. If the prey population has a growth law that leads to a stable carrying capacity K, we do not expect the predator population to grow extinct near K since it has lots of little creatures to munch upon[1]. This suggests that an equilibrium point $(K, 0)$ is another saddle.

Given the above, we can already construct the phase plane portraits

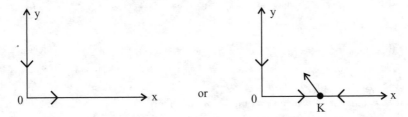

and now all that remains is to analyze the stability of any other equilibria. The ever popular model is the Lotka–Volterra model whose equations are

$$\dot{x} = rx - \alpha xy, \quad \dot{y} = -sy + \beta xy$$

where all parameters are positive. Then X grows exponentially in the absence of Y and Y goes extinct in the absence of X. The equilibrium point $(s/\beta, r/\alpha)$ is a center for the linearized model and it is preserved in the full model. This leads to the wonderful harmonious universe phase plane portrait of ovoid shaped periodic solutions filling the region $x > 0$, $y > 0$.

The author has always been bothered by this model, picturing a situation where there are two rabbits quivering with fear and 10^n coyotes—not to worry things will improve. A more tragic scenario would be if both rabbits were of the same sex. But setting flippancy aside, the Lotka–Volterra model lacks a requirement of a good model—*structural stability*. This can be loosely interpreted as robustness of the model under small perturbations.

For instance, modify the equation for X assuming that in the absence of Y its growth is logistic with a very large carrying capacity $K = 1/\epsilon$, $\epsilon > 0$ and small. One obtains

---

[1] However, an amusing possibility would be that if the prey population gets big enough it starts destroying predators (the killer rabbits model). Then $(K, 0)$ would be a stable node.

the equation for X

$$\dot{x} = rx(1 - \epsilon x) - \alpha xy$$
$$= rx - \alpha xy - \epsilon rx^2,$$

which is certainly a small perturbation of the original equation. The new equilibrium point (x_∞, y_∞) will be $(s/\beta, r/\alpha - \epsilon s/\beta\alpha)$ which will be as close as we want to $(s/\beta, r/\alpha)$ if ϵ is small enough. But a little analysis shows that it will be a stable spiral or a stable node for that range of ϵ so the periodicity is lost. An interesting discussion of further variants of the Lotka–Volterra model can be found in the book by Polking, Boggess, and Arnold.

5   Harvesting

If we consider the standard competition model with logistic growth for both populations, and where their per capita rate of growth is reduced proportionally to the number of competitors present, we have the model

$$\dot{x} = rx \left(1 - \frac{x}{K}\right) - \alpha xy$$
$$= x(r - rx/K - \alpha y),$$

$$\dot{y} = sy \left(1 - \frac{y}{J}\right) - \beta xy$$
$$= y(s - sy/J - \beta x),$$

where all parameters are positive. The linear curves in parentheses, called nullclines, are graphed and we get the possible pictures

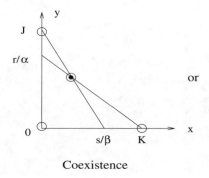

Coexistence

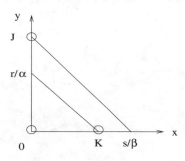

Competitive Exclusion

where the equilibria are circled.

Now suppose the population Y is harvested at a constant rate $H > 0$ caused by hunting, fishing, or disease, for instance. The second equation becomes

$$\dot{y} = y(s - sy/J - \beta x) - H$$

and the right-hand side is a hyperbolic curve whose maximum moves downward as H is increased. The successive pictures in the coexistence case look like

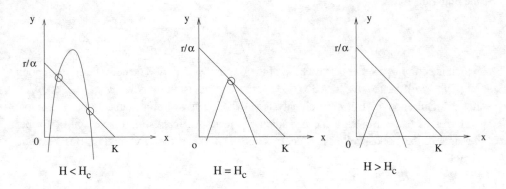

$$H < H_c \qquad\qquad\qquad H = H_c \qquad\qquad\qquad H > H_c$$

and note that the equilibria $(0,0)$ and $(K,0)$ have moved below the x-axis. Along and near the x-axis we have $\frac{dy}{dt} \approx -H < 0$ so that once there are no equilibria in the positive xy-quadrant, then X wins all the marbles (or birdseed). The model is quite similar to that for the one-population logistic model with harvesting discussed in Chapter 2.

The critical value H_c of the harvest rate is easily found by solving

$$r - \frac{rx}{K} - \alpha y = 0$$

for x, then substituting it into

$$y(s - sy/J - \beta x) - H = 0$$

and solving the resulting quadratic equation. The value of H beyond which there will not be real roots will be H_c.

Example. A system with coexistence is

$$\dot{x} = x\left[\frac{1}{2} - \frac{1}{400}x - \frac{1}{10^3}y\right], \qquad \dot{y} = y\left[\frac{4}{5} - \frac{1}{500}y - \frac{1}{10^3}x\right],$$

with saddle points $(200,0)$ and $(0,400)$, and $(x_\infty, y_\infty) = (50, 375)$ a stable node. To determine the critical harvest $H = H_c$ when the y-population is harvested, first solve for x:

$$\frac{1}{2} - \frac{1}{400}x - \frac{1}{10^3}y = 0 \Rightarrow x = 200 - \frac{2}{5}y.$$

Next solve

$$y\left[\frac{4}{5} - \frac{1}{500}y - \frac{1}{10^3}\left(200 - \frac{2}{5}y\right)\right] - H = 0$$

which simplifies to $2y^2 - 750y + 1250H = 0$. Its roots are

$$y = \frac{1}{4}\left(750 \pm \sqrt{750^2 - 10^4 H}\right)$$

and they will no longer be real when

$$10^4 H > 750^2 \quad \text{or} \quad H > 750^2/10^4 = 56.25 = H_c.$$

One can apply harvesting to competitive exclusion cases and create coexistence, as well as to predator-prey models to destabilize them. More complicated models than the logistic type ones could make interesting projects, possibly leading one to drop mathematics altogether and take up wildlife management. Happy Harvesting!

6 A Conservative Detour

A pleasant respite from the onerous task of solving nonlinear equations in several variables to find equilibria, then linearizing, calculating eigenvalues, etc., is afforded by examining *conservative systems*. No knowledge of political science is required, and the analysis of the nature of equilibria in the two-dimensional case is simply a matter of graphing a function of one variable and locating its maxima, minima, or inflection points—a strictly calculus exercise. Since the differential equations model linear or nonlinear, undamped, mass-spring systems or pendulums, one might consider briefly studying them prior to embarking on the discussion of almost linear systems.

The standard equation for a one degree of freedom conservative system is

$$\ddot{x} + f(x) = 0,$$

where $x = x(t)$ is a scalar function. This could describe the motion of a particle along a line or curve, where the restoring force $f(x)$ is only dependent on the displacement, so there is *no damping*, i.e. no $\dot{x}$ term present. Some simple examples are $f(x) = k^2 x$, the undamped oscillator, $f(x) = \frac{g}{L}\sin x$, the pendulum, or $f(x) = x + x^3$, a nonlinear spring.

We will assume that $f(x)$ is sufficiently nice so that its antiderivative $F(x) = \int^x f(s)ds$ is defined and satisfies $F'(x) = f(x)$. Now multiply the differential equation by $\dot{x}$

$$\ddot{x}\dot{x} + f(x)\dot{x} = 0,$$

then integrate to obtain the *energy equation*:

$$\frac{\dot{x}^2}{2} + F(x) = C, \text{ a constant.}$$

This is the key equation; the first term on the left is a measure of the kinetic energy per unit mass, and the second is a measure of the potential energy per unit mass. Consequently, the energy equation says that along a solution

$$(\text{kinetic energy}) + (\text{potential energy}) = \text{constant},$$

which is the defining characteristic of conservative systems—their total energy is constant along trajectories.

Moving to the phase plane, where $(x, \dot{x}) = (x, y)$, the energy equation becomes

$$E(x, y) = \frac{1}{2}y^2 + F(x) = C,$$

and we can define $z = E(x, y)$ as the *energy surface*. Its level curves $E(x, y) = C$, in the phase plane, are therefore the trajectories of the two-dimensional system

$$\dot{x} = y, \quad \dot{y} = -f(x).$$

For the simple oscillator $f(x) = x$, so $F(x) = \frac{1}{2}x^2$, and the energy surface is the paraboloid $z = \frac{1}{2}y^2 + \frac{1}{2}x^2$ whose level curves are circles $\frac{1}{2}y^2 + \frac{1}{2}x^2 = C$, reflecting the fact that the equilibrium point (0,0) is a center. The diagram below shows the values of the total energy $C = K_1, K_2$.

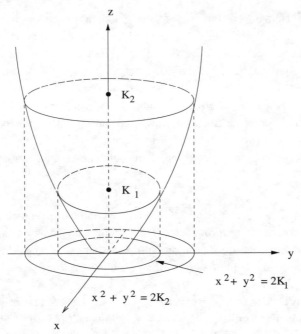

The equilibria of a conservative system are points $(\bar{x}, 0)$ where $f(\bar{x}) = 0$, and since $F'(x) = f(x)$ they are *critical points* of $F(x)$; this is the key to the analysis. Rewrite the energy equation as

$$\frac{1}{2}y^2 = C - F(x) \quad \text{or} \quad y = \pm\sqrt{2}\,\sqrt{C - F(x)},$$

and then compute the values of y near $x = \bar{x}$ for selected values of the total energy C. For each x and C there will be two such values, in view of the $\pm$ sign, which tells us that trajectories will be symmetric about the x-axis.

For gorgeous, exact graphs this plotting can be done using graphing technology, but to sketch trajectories and analyze the stability of $(\bar{x}, 0)$, we only need eyeball estimates of the quantity $\sqrt{C - F(x)}$ near the critical point $\bar{x}$. The various possibilities are:

(i) $F(\bar{x})$ is a local maximum. Plot $z = F(x)$ near $x = \bar{x}$ then select values $z = K_1, K_2, \ldots,$ of the total energy and intersect the horizontal lines $z = K_i$ with the graph. Align an (x, y) graph below the (x, z) graph then plot or estimate the points $(x, \pm\sqrt{2}\,\sqrt{K_i - F(x)})$. We see that $(\bar{x}, 0)$ is a saddle point.

(ii) $F(\bar{x})$ is a local minimum. This case is sometimes referred to as a *potential well* or *sink*. Proceed as in i) and we see that $(\bar{x}, 0)$ is a center.

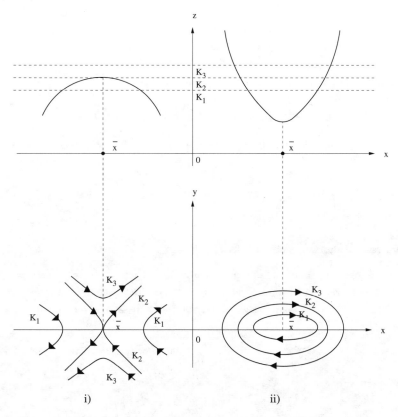

i)   ii)

(iii) $(\bar{x}, F(\bar{x}))$ is a point of inflection. The trajectory through $(\bar{x}, 0)$ will have a cusp and it is sometimes called a degenerate saddle point—it is unstable.

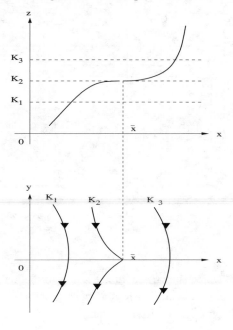

Since i), ii) and iii) are the only possibilities we can conclude that conservative systems $\ddot{x} + f(x) = 0$ can only have equilibria which are saddle points (unstable) or centers (stable), and there will be no asymptotically stable equilibria.

To plot a specific trajectory, given initial conditions

$$\big(x(t_0), \dot{x}(t_0)\big) = \big(x(t_0), y(t_0)\big) = (x_0, y_0),$$

use the fact that its total energy is constant. Hence

$$E(x, y) = \frac{1}{2}y^2 + F(x) = \frac{1}{2}y_0^2 + F(x_0)$$

is the required implicit representation.

Examples. We start off with the classic

1. $\ddot{x} + \frac{g}{L}\sin x = 0$, the undamped pendulum. Then $f(x) = \frac{g}{L}\sin x$ so $F(x) = -\frac{g}{L}\cos x$ which has local minima at $\bar{x} = \pm 2n\pi$ and local maxima at $\bar{x} = \pm(2n+1)\pi$, $n = 0, 1, 2, \dots$. Given total energy C, trajectories will satisfy the equation

$$y = \pm\sqrt{2}\sqrt{C + \frac{g}{L}\cos x},$$

and the picture is:

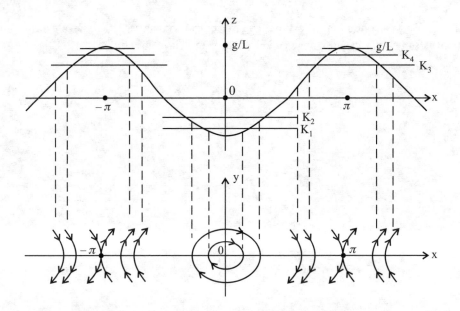

This is a local picture which shows that a moderate perturbation from rest $(x = y = 0)$ will result in a stable, periodic orbit, whereas a small perturbation from $x = \pm\pi$, $y = 0$, the pendulum balanced on end, results in instability. A more complete picture of the phase plane looks like this

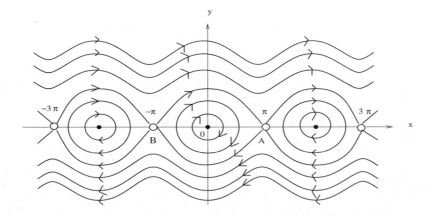

The unbounded trajectories at the top and bottom of the graph represent trajectories whose energy is so large that the pendulum never comes to rest. For $0 < C < \frac{g}{L}$ there is a potential well centered at $(0, 0)$, which implies that initial conditions

$$|x(0)| < \pi, \quad |\dot{x}(0)| = |y(0)| < \sqrt{2}\sqrt{\frac{g}{L} + \frac{g}{L}\cos x(0)},$$

will give a periodic orbit around the origin.

Remark. A potential well has a nice physical interpretation if we think of a bead sliding along a frictionless wire.

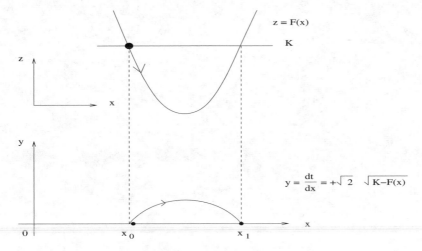

If the bead is started at the point $(x_0, K) = (x_0, F(x_0))$ in the xz-plane, corresponding to the point $(x_0, 0)$ in the xy-plane, its velocity and kinetic energy are zero, and its potential energy is $F(x_0)$. As the bead slides down the wire its kinetic energy increases, reaches a maximum, then returns to zero at the point $(x_1, K) = (x_1, F(x_1))$ corresponding to the point $(x_1, 0)$ in the xy-plane. There it will reverse direction $\left(y = \frac{dx}{dt} < 0\right)$ and return to the point (x_0, K).

2. $\ddot{x} + \frac{g}{L}\left(x - \frac{x^3}{6}\right) = 0$, the approximate pendulum with $\sin x$ replaced by the approximation $x - \frac{x^3}{6}$, valid for small $|x|$. Then

$$f(x) = \frac{g}{L}\left(x - \frac{x^3}{6}\right) \quad \text{and} \quad F(x) = \frac{g}{2L}\left(x^2 - \frac{x^4}{12}\right)$$

which has a local minimum at $x = 0$, and maxima at $x = \pm\sqrt{6}$, with $F(\pm\sqrt{6}) = 3g/2L$. Periodic motions will exist for

$$|x(0)| < \sqrt{6}, \qquad |\dot{x}(0)| = |y(0)| < \sqrt{\frac{g}{L}}\sqrt{3 - x(0)^2 + \frac{x^4(0)}{12}},$$

and the picture is the following one:

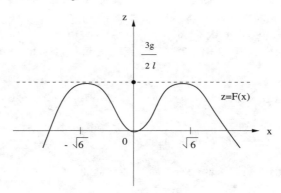

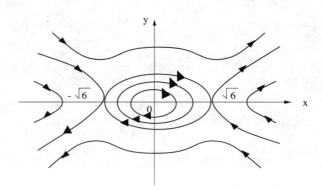

3. $\ddot{x} + x + x^2 = 0$, so

$$f(x) = x + x^2, \quad \text{and} \quad F(x) = \frac{x^2}{2} + \frac{x^3}{3}$$

which has a local minimum at $x = 0$, a local maximum at $x = -1$, and $F(-1) = 1/6$. There is a potential well at $x = 0$ but it will not be symmetric as in the previous two examples.

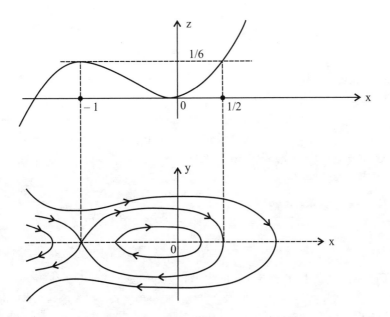

The reader might want to look at examples like

a) $f(x) = \frac{x}{x-1}$,

b) $f(x) = x - 1 - \lambda x^2$, $0 < \lambda < 1/4$, but examine the case $\frac{1}{4} \le \lambda$,

c) $f(x) = \begin{cases} -1 & \text{if } x \le -1 \\ 0 & \text{if } |x| < 1 \\ 1 & \text{if } x \ge 1. \end{cases}$

An important question is the following one: given a closed orbit of a nonlinear system, representing a periodic solution, what is its period T? For most autonomous systems this is a difficult analytic/numeric question. If one is able to obtain a point (x_0, y_0) on the orbit, then one can numerically chase around the orbit and hope to return close to the initial point, and then estimate the time it took. If the orbit is an isolated limit cycle even finding an (x_0, y_0) may be a formidable task.

But for two-dimensional conservative systems it is easy to find an expression for the period. We will assume for simplicity that the orbit is symmetric with respect to the y-axis so it looks like this:

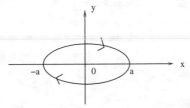

If the equation is $\ddot{x} + f(x) = 0$, then when $x = a$, $\dot{x} = y = 0$ so the energy equation is

$$E(x, y) = E(x, \dot{x}) = \frac{\dot{x}^2}{2} + F(x) = F(a), \quad F'(x) = f(x),$$

or

$$\frac{dx}{dt} = \sqrt{2} \, (F(a) - F(x))^{1/2}.$$

We have used the $+$ sign since we will compute in the region $\dot{x} = y \geq 0$. The last equation implies that

$$dt = \frac{1}{\sqrt{2} \, (F(a) - F(x))^{1/2}} \, dx,$$

and if we integrate along the orbit from $x = 0$ to $x = a$ we get 1/4 of the time it takes to traverse the orbit, hence

$$\text{Period} = T(a) = 4 \int_0^a \frac{1}{\sqrt{2} \, (F(a) - F(x))^{1/2}} \, dx$$

$$= 2\sqrt{2} \int_0^a \frac{1}{(F(a) - F(x))^{1/2}} \, dx.$$

The expression for the period is a lovely one, but has a slightly venomous bite since the integral is improper because the denominator vanishes at $x = a$. Nevertheless, the period is finite so the integral must be finite—with care it can be evaluated numerically.

Let's look at the pendulum equation

$$\ddot{x} + \frac{g}{L} \sin x = 0, \quad x(0) = a, \quad \dot{x}(0) = 0,$$

and some approximations:

1st Approximation: $\sin x \approx x$ and we'll play dumb and suppose we don't know how to solve $\ddot{x} + \frac{g}{L} x = 0$. Then $F(x) = \frac{g}{L} \frac{x^2}{2}$ and

$$T(a) = 2\sqrt{2} \int_0^a \frac{1}{\sqrt{\frac{g}{L} \left(\frac{a^2}{2} - \frac{x^2}{2} \right)^{1/2}}} \, dx = 4\sqrt{\frac{L}{g}} \sin^{-1} \frac{x}{a} \Big]_0^a = 2\pi \sqrt{\frac{L}{g}}$$

which is independent of a and valid only for small $|x|$. If $L = 1$m. and $g = 9.8$ m/sec^2 then $T \approx 2.007090$.

2nd Approximation: $\sin x \approx x - \frac{x^3}{6}$, then $F(x) = \frac{g}{L} \left(\frac{x^2}{2} - \frac{x^4}{24} \right)$ and

$$T(a) = 2\sqrt{2} \int_0^a \frac{1}{\sqrt{\frac{g}{L} \left(\frac{a^2}{2} - \frac{a^4}{24} - \frac{x^2}{2} + \frac{x^4}{24} \right)}} \, dx.$$

We computed it for $a = 1/2, 1, 3/2$; the second column is the value for $L = 1$m. and $g = 9.8$ m/sec^2.

$$T\left(\frac{1}{2}\right) = 2\sqrt{2} \, \sqrt{\frac{L}{g}} (2.257036) \qquad 2.039250$$

$$T(1) = 2\sqrt{2} \, \sqrt{\frac{L}{g}} (2.375832) \qquad 2.146583$$

$$T(3/2) = 2\sqrt{2} \, \sqrt{\frac{L}{g}} (1.628707) \qquad 2.375058$$

3rd Approximation: the full pendulum so $F(x) = -\frac{g}{L}\cos x$ and

$$T(a) = 2\sqrt{2} \int_0^a \frac{1}{\sqrt{\frac{g}{L}}(\cos x - \cos a)^{1/2}}\,dx.$$

Then as above

$$T\left(\frac{1}{2}\right) = 2\sqrt{2}\,\sqrt{\frac{L}{g}}(2.256657) \qquad 2.038907$$

$$T(1) = 2\sqrt{2}\,\sqrt{\frac{L}{g}}(2.368799) \qquad 2.140229$$

$$T(3/2) = 2\sqrt{2}\,\sqrt{\frac{L}{g}}(2.584125) \qquad 2.334777$$

For a large class of conservative systems there is a direct relation of solutions and periods to Jacobi elliptic functions and elliptic integrals. The article by Kenneth Meyer gives a nice discussion of this.

If the reader is in a number crunching mood he or she might want to look at:

a) $f(x) = x^3$, with $x(0) = a, \dot{x}(0) = 0; a = 1, a = 24$.

b) The step function $f(x)$ given in c) above and determine the minimum period of oscillation.

c) The example $f(x) = x + x^2$ given previously; the limits of integration in the period integral must be adjusted since the orbit is not symmetric in the x-direction.

d) Compare the approximate and full pendulum's period when a is close to π, say $a = 31/10$, so the pendulum is nearly vertical.

Finally, we want to briefly look at the effect of damping on a conservative system. When we examined the second order linear equations

$$\ddot{x} + \omega^2 x = 0 \quad \text{and} \quad \ddot{x} + c\dot{x} + \omega^2 x = 0, \quad c > 0,$$

we saw that the addition of the damping term $c\dot{x}$ resulted in all solutions approaching zero as $t \to \infty$. In the phase plane the origin, which was a stable center in the first equation, became an asymptotically stable spiral point or node. A simple phase plane analysis for the damped pendulum

$$\ddot{x} + c\dot{x} + \frac{g}{L}\sin x = 0, \qquad 0 < c < 2\sqrt{g/L},$$

shows that the saddle points of the undamped pendulum are sustained, whereas the centers become asymptotically stable spiral points.

For the conservative system $\ddot{x} + f(x) = 0$ we can model the addition of damping by considering the general equation

$$\ddot{x} + g(x, \dot{x})\dot{x} + f(x) = 0,$$

where the middle term represents the effect of damping—$g(x, \dot{x})$ could be constant as in the case above. Multiply the equation by $\dot{x}$ and transpose terms to obtain

$$\ddot{x}\dot{x} + f(x)\dot{x} = -g(x, \dot{x})\dot{x}^2,$$

and if $F'(x) = f(x)$ this can be written as

$$\frac{d}{dt}\left(\frac{\dot{x}^2}{2} + F(x)\right) = -g(x, \dot{x})\dot{x}^2.$$

But the expression in parentheses is the energy equation, and if we let $\dot{x} = y$ then we obtain the equation

$$\frac{d}{dt}E(x, y) = -g(x, y)y^2$$

which gives us a measure of the rate of growth of the energy of the system.

For instance, if $g(x, y) > 0$ for all x, y, the case of *positive damping*, e.g. $g(x, y) = c > 0$, then the energy of the system is strictly decreasing, so any nontrivial motion eventually ceases—the system is dissipative.

Example: For any conservative system $\ddot{x} + f(x) = 0$ the addition of a damping term $c\dot{x}$, $c > 0$, results in the energy growth equation

$$\frac{d}{dt}E(x, y) = -cy^2$$

so all motions eventually damp out. Similarly for the equation $\ddot{x} + |\dot{x}|\dot{x} + f(x) = 0$, we have

$$\frac{d}{dt}E(x, y) = -|y|y^2$$

and the behavior is the same.

On the other hand if $g(x, y) < 0$ for all x, y, the case of *negative damping*, then the energy of the system is increasing. The effect is that of an internal source pumping in energy.

But the most interesting case is one where both kinds of damping occur in different regions of the phase space, and the balance between energy loss and gain can result in *self-sustained oscillations* or limit cycles. A simple example is the following one:

$$\ddot{x} + (x^2 + \dot{x}^2 - 1)\dot{x} + x = 0,$$

and the energy growth equation is

$$\frac{d}{dt}E(x, y) = -(x^2 + y^2 - 1)y^2.$$

When $x^2 + y^2 < 1$ the damping is negative, so energy increases, whereas when $x^2 + y^2 > 1$, the damping is positive and energy decreases. It is easily seen that $x(t) = \cos t$ is a solution, which gives an isolated closed trajectory $x^2 + y^2 = 1$ in the phase plane, hence a limit cycle.

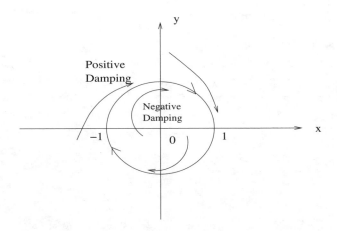

The extensively studied Van der Pol equation

$$\ddot{x} + \epsilon(x^2 - 1)\dot{x} + x = 0, \quad \epsilon > 0,$$

has the energy growth equation

$$\frac{d}{dt}E(x, y) = -\epsilon(x^2 - 1)y^2.$$

For $|x| < 1$ energy is increasing, whereas for $|x| > 1$ it is decreasing, so we would expect to find self-sustained oscillations. It has been shown that for *any* $\epsilon > 0$ there exists a stable limit cycle; the proof is complicated but elegant. For $\epsilon < 0.1$ it is closely approximated by $x(t) = 2\cos t$, hence is circular in shape, with period $T \cong 2\pi$, whereas for $\epsilon \gg 1$ it becomes quite jerky and is an example of a *relaxation oscillation*.

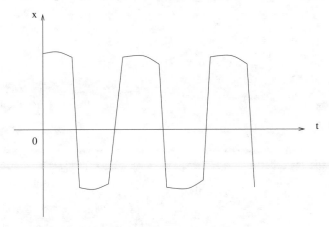

Enough! The author's detour has arrived at a point where it is best to return to the main highway, and leave the reader to further explore the fascinating territory just briefly described.

7 Stability and Gronwall-ing

Preliminaries Up to this point, except for several nonautonomous first order equations in Chapter 2, the discussion of stability of solutions has been limited to an analysis of the solution $x(t) \equiv 0$ of constant coefficient second order equations, and the equilibrium point (0,0) of the system $\dot{x} = Ax$, where A is a 2×2 constant matrix. We want to move a little further afield and to do so we need a measure of distance or a norm in R^n.

First of all if $x = \operatorname{col}(x_1, \ldots, x_n)$ is a point (vector) in R^n then define the norm of x by

$$\|x\| = \sum_1^n |x_i|.$$

This is not the Euclidean norm, but is equivalent, is easier to work with, and has all the required properties. Then we need a measure of the size of a matrix, so if $A = (a_{ij})$, $i, j = 1, \ldots, n$, is an $n \times n$ matrix we define the norm of A by

$$\|A\| = \sum_{i,j=1}^n |a_{ij}|,$$

and it has the properties we need:

$\|A + B\| \le \|A\| + \|B\|, \|AB\| \le \|A\|\|B\|, \|cA\| \le |c|\|A\|$ for any scalar c, $\|Ax\| \le \|A\|\|x\|$ for any vector x.

Obviously, if $A = A(t)$ is continuous, i.e. has continuous entries $a_{ij}(t)$, then $\|A(t)\|$ is continuous.

To precisely define the notion of stability and asymptotic stability of a solution $x(t)$, defined for $t_0 \le t < \infty$, would require some delta-epsilonics, which the reader can find in any intermediate or advanced text. The following picture describes stability of the solution $x(t) = x_0$; we have selected a constant solution for artistic simplicity.

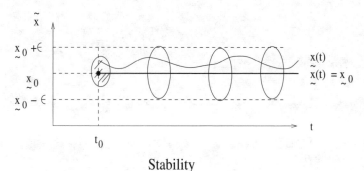

Stability

What the picture means is that given an arbitrary $\epsilon > 0$ we can construct a "tube" of radius ϵ around $x(t) = x_0$, and any solution $x(t)$ satisfying $x(t_0) = x_1$, where x_1 is in a smaller shaded disc centered at (t_0, x_0), will be defined and stay in the tube for all $t \ge t_0$. Once close—it stays close.

The above definition does not prevent a solution from entering the tube at a later time than t_0 then leaving it. To preclude this we need the much stronger notion of uniform stability which we will not discuss. For asymptotic stability of the solution $\underset{\sim}{x}(t) = \underset{\sim}{x_0}$ the picture is this one:

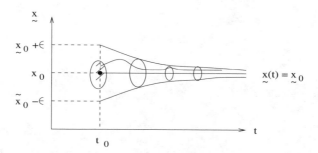

Asymptotic Stability

We can say that $\underset{\sim}{x}(t)$ once close—it stays close, and furthermore the diameter of the tube goes to zero as $t \to \infty$, so

$$\lim_{t \to \infty} \underset{\sim}{x}(t) = \underset{\sim}{x_0}.$$

The reader's mind's eye can now substitute a nonconstant solution $\underset{\sim}{x}(t)$ satisfying $\underset{\sim}{x}(t_0) = x_0$, and construct a wiggly tube around it to infer the notion of what is meant by stability and asymptotic stability of the solution $\underset{\sim}{x}(t)$.

Stability for Linear Systems

For the linear systems $\dot{\underset{\sim}{x}} = A\underset{\sim}{x}$, $A = (a_{ij})$, $i, j = 1, \ldots, n$, or $\dot{\underset{\sim}{x}} = A(t)\underset{\sim}{x}$, $A(t) = (a_{ij}(t))$, $i, j = 1, \ldots, n$, where each $a_{ij}(t)$ is continuous for $t \geq t_0$, we know the solution $\underset{\sim}{x}(t)$ of the IVP

$(\star)$ $\qquad\qquad \dot{\underset{\sim}{x}} = A\underset{\sim}{x} \quad \text{or} \quad \dot{\underset{\sim}{x}} = A(t)\underset{\sim}{x}, \quad \underset{\sim}{x}(t_0) = \underset{\sim}{x_0},$

is given via the fundamental matrix $\Phi(t) = (\varphi_{ij}(t))$,

$$\underset{\sim}{x}(t) = \Phi(t)\underset{\sim}{x_0}, \quad \Phi(t_0) = I,$$

and the columns of $\Phi(t)$ are a fundamental system of solutions.

From the last statement we can imply several things about $\|\Phi(t)\|$:

(i) If every solution of $(\star)$ is bounded for $t \geq t_0$, that means that every entry $\varphi_{ij}(t)$ is bounded, so there exists some constant $M > 0$ such that $\|\Phi(t)\| \leq M$ for $t \geq t_0$.

(ii) In the case of $A(t) = A$ suppose every eigenvalue of A has a negative real part. A little cogitation of what this implies about each entry $\varphi_{ij}(t)$ will convince one that there exists constants $R > 0$, $\alpha > 0$ such that $\|\Phi(t)\| \leq Re^{-\alpha t}$ for $t \geq t_0 \geq 0$.

Example. $A = \left(\begin{smallmatrix} -4 & -1 \\ 1 & -2 \end{smallmatrix} \right)$, a fundamental matrix is $\left(\begin{smallmatrix} e^{-3t} & -e^{-3t}+te^{-3t} \\ -e^{-3t} & -te^{-3t} \end{smallmatrix} \right) = \Omega(t)$.
Multiply by $\Omega(0)^{-1} = \left(\begin{smallmatrix} 0 & -1 \\ -1 & -1 \end{smallmatrix} \right)$ to get $\Phi(t)$ satisfying $\Phi(0) = I$,

$$\Phi(t) = \begin{pmatrix} e^{-3t} - te^{-3t} & -te^{-3t} \\ te^{-3t} & e^{-3t} + te^{-3t} \end{pmatrix}.$$

Then

$$\|\Phi(t)\| = |e^{-3t} - te^{-3t}| + |te^{-3t}| + |-te^{-3t}| + |e^{-3t} + te^{-3t}| \leq 2e^{-3t} + 4te^{-3t}.$$

A little analysis shows that if $0 \leq \alpha < 2.63$ then $e^{-\alpha t} > te^{-3t}$ for $t > 0$, so let $\alpha = 1$
then

$$\|\Phi(t)\| \leq 2e^{-3t} + 4e^{-t} \leq 6e^{-t}.$$

The first case of stability to logically consider is the general one $A = A(t)$, but we must know the structure of $\Phi(t)$ which will usually be very difficult to find. Here's a sometimes helpful result:

If all solutions of $\dot{x} = A(t)x$, $t \geq t_0$, are bounded then they are stable.

Let $x_1(t)$ be the solution satisfying $x_1(t_0) = x_0$, and $x_2(t)$ be the solution satisfying $x_2(t_0) = x_1$. Since all solutions are bounded we know $\|\Phi(t)\| \leq M$, so let $\|x_0 - x_1\| < \epsilon/M$ and the simple estimate

$$\|x_2(t) - x_1(t)\| = \|\Phi(t)(x_1 - x_0)\| \leq \|\Phi(t)\|\|x_1 - x_0\| \leq M\|x_1 - x_0\| < \epsilon$$

gives the desired conclusion. For a constant matrix A things get better:

If all the eigenvalues of A have negative real parts then every solution of $\dot{x} = Ax$ is asymptotically stable.

Just replace the constant M in the previous estimate with $Re^{-\alpha t}$, $R > 0$, $\alpha > 0$ to obtain

$$\|x_2(t) - x_1(t)\| \leq Re^{-\alpha t}\|x_1 - x_0\|.$$

This shows that solutions are bounded since $\|x(t)\| \leq Re^{-\alpha t}\|x_0\|$ and furthermore $\lim_{t \to \infty}(x_2(t) - x_1(t)) = 0$.

Way back in Chapter 2 we introduced Gronwall's Lemma and touted it as a very useful tool to study stability. To live up to our claim we examine the nonautonomous system

$$\dot{x} = (A + C(t))x, \quad x(0) = 0; \quad C(t) = (c_{ij}(t)), \qquad i, j = 1, \ldots, n,$$

where each $c_{ij}(t)$ is continuous for $0 \leq t < \infty$. Our intuition would tell us that if $C(t)$ were "small" in some sense, the behavior of solutions would mimic those of the system $\dot{x} = Ax$. We follow that line of reasoning.

First, recall that for the general IVP

$$\dot{x} = Ax + B(t), \quad x(0) = x_0$$

the solution is given by the variation of parameters formula

$$\underset{\sim}{x}(t) = \Phi(t)\underset{\sim}{x_0} + \int_0^t \Phi(t-s)\underset{\sim}{B}(s)ds.$$

The trick is to rewrite the system $\underset{\sim}{\dot{x}} = (A + C(t))\underset{\sim}{x}$ as $\underset{\sim}{\dot{x}} = A\underset{\sim}{x} + C(t)\underset{\sim}{x}$ and let the last term be our $\underset{\sim}{B}(t)$. Then we can express its solution as the integral equation

$$\underset{\sim}{x}(t) = \Phi(t)\underset{\sim}{x_0} + \int_0^t \Phi(t-s)C(s)\underset{\sim}{x}(s)ds$$

which is ripe for Gronwall-ing.

Suppose all the eigenvalues of A have negative real parts, then $\|\Phi(t)\| \le Re^{-\alpha t}$, for some positive α and R and $t \ge 0$. Take norms in the previous equation

$$\|\underset{\sim}{x}(t)\| \le \|\Phi(t)\|\|\underset{\sim}{x_0}\| + \int_0^t \|\Phi(t-s)\|\|C(s)\|\|\underset{\sim}{x}(s)\|ds$$

$$\le Re^{-\alpha t}\|\underset{\sim}{x_0}\| + \int_0^t Re^{-\alpha(t-s)}\|C(s)\|\|\underset{\sim}{x}(s)\|ds,$$

and multiplying through by $e^{\alpha t}$ gives

$$\|\underset{\sim}{x}(t)\|e^{\alpha t} \le R\|\underset{\sim}{x_0}\| + \int_0^t R\|C(s)\|\|\underset{\sim}{x}(s)\|e^{\alpha s}ds.$$

Now apply Gronwall's Lemma

$$\|\underset{\sim}{x}(t)\|e^{\alpha t} \le R\|\underset{\sim}{x_0}\| \exp\left[R\int_0^t \|C(s)\|ds\right]$$

$$\le R\|\underset{\sim}{x_0}\| \exp\left[R\int_0^\infty \|C(s)\|ds\right],$$

or

$$\|\underset{\sim}{x}(t)\| \le R\|\underset{\sim}{x_0}\|e^{-\alpha t} \exp\left[R\int_0^\infty \|C(s)\|ds\right].$$

We conclude that if the integral $\int_0^\infty \|C(s)\|ds$ is finite then

i) All solutions are bounded, hence stable,

ii) All solutions approach zero as $t \to \infty$ since $\alpha > 0$.

But the system is linear and homogeneous, so the difference of any two solutions is a solution. We can conclude

> Given $\underset{\sim}{\dot{x}} = (A + C(t))\underset{\sim}{x}$, where $C(t)$ is a continuous matrix for $0 \le t < \infty$. If the eigenvalues of A have negative real parts and $\int_0^\infty \|C(t)\|dt < \infty$, then all solutions are asymptotically stable.

Example. $\ddot{x} + \left(2 + \frac{1}{1+t^2}\right)\dot{x} + x = 0$ can be expressed as the system

$$\begin{pmatrix}\dot{x} \\ \dot{y}\end{pmatrix} = \begin{pmatrix} 0 & 1 \\ -1 & -2-\frac{1}{1+t^2}\end{pmatrix}\begin{pmatrix}x \\ y\end{pmatrix} = \left[\begin{pmatrix} 0 & 1 \\ -1 & -2\end{pmatrix} + \begin{pmatrix} 0 & 0 \\ 0 & \frac{-1}{1+t^2}\end{pmatrix}\right]\begin{pmatrix}x \\ y\end{pmatrix}.$$

The matrix A has a double eigenvalue

$$\lambda = -1, \quad \text{and} \quad \int_0^\infty \frac{1}{1+t^2}\, dt$$

is finite so all solutions are asymptotically stable.

The conditions on the growth of $C(t)$ can be weakened, but the intent was to demonstrate a simple application of Gronwall's Lemma, not give the definitive result.

Stability for Nonlinear Systems

At this point in the development of potential ODE practitioners, their only likely exposure to nonlinear systems are to those sometimes referred to as almost linear systems. They may come in different guises such as conservative or Hamiltonian systems, but the basic structure is the following one, where we stick with $n = 2$ for simplicity:

Given the autonomous system

$$\dot{x} = P(x,y), \quad \dot{y} = Q(x,y),$$

where P and Q are smooth functions, and (0,0) is an isolated equilibrium point. Then since $P(0,0) = Q(0,0) = 0$ we can apply Taylor's Theorem and write

(L)
$$\dot{x} = P_x(0,0)x + P_y(0,0)y + \left(\begin{matrix} \text{higher order terms} \\ \text{in } x \text{ and } y \end{matrix}\right)$$

$$\dot{y} = Q_x(0,0)x + Q_y(0,0)y + \left(\begin{matrix} \text{higher order terms} \\ \text{in } x \text{ and } y \end{matrix}\right).$$

The higher order terms will be second degree or higher polynomials in x and y. If the equilibrium point is $(x_0, y_0) \neq (0,0)$ we can make the change of variables

$$x = u + x_0, \quad y = v + y_0,$$

and this will convert the system (L) to one in u and v with $(0,0)$ an equilibrium point; the first partial derivatives will be evaluated at (x_0, y_0). The system (L), *neglecting* the higher order terms, is the linearized system corresponding to the original system, and we use it to study the stability of (0,0).

If we let $\underset{\sim}{x} = \text{col}(x_1, \ldots, x_n)$ we can write a more general form of (L):

$$\dot{\underset{\sim}{x}} = A\underset{\sim}{x} + f(\underset{\sim}{x})$$

where A is a constant $n \times n$ matrix and $f(\underset{\sim}{x})$ is higher order terms, which may be infinite in number, as in the Taylor series expansion. What is important is that $f(\underset{\sim}{x})$ satisfies

$$\|f(\underset{\sim}{x})\|/\|\underset{\sim}{x}\| \to 0 \quad \text{as} \quad \|\underset{\sim}{x}\| \to 0$$

and this relation is independent of the choice of norm. Systems of this form are called *almost linear systems.*

Examples.

a) Suppose the higher order terms (H.O.T. if you want to appear hip) are $x^2 + 2xy + y^4$. Since

$$\frac{1}{|x| + |y|} \leq \begin{cases} 1/|x| \\ 1/|y| \end{cases}$$

then

$$\frac{\|f(\underset{\sim}{x})\|}{\|\underset{\sim}{x}\|} = \frac{|x^2 + 2xy + y^4|}{|x| + |y|} \leq \frac{|x|^2 + 2|x||y| + |y|^4}{|x| + |y|}$$

$$\leq \frac{|x|^2}{|x|} + \frac{2|x||y|}{|y|} + \frac{|y|^4}{|y|} = 3|x| + |y|^3$$

which approaches 0 as $|x| + |y| \to 0$.

b) But one could use the Euclidean norm $r = \sqrt{x^2 + y^2} = \|\underset{\sim}{x}\|$, then convert to polar coordinates, $x = r\cos\theta$, $y = r\sin\theta$, to obtain

$$\frac{\|f(\underset{\sim}{x})\|}{\|\underset{\sim}{x}\|} = \frac{|r^2\cos^2\theta + 2r^2\cos\theta\sin\theta + r^4\sin^4\theta|}{r}$$

$$\leq \frac{r^2 + 2r^2 + r^4}{r} = 3r + r^3 \to 0 \quad \text{as} \quad r \to 0.$$

c) The undamped pendulum $\ddot{x} + \frac{g}{L}\sin x = 0$ can be written as the system

$$\dot{x} = y, \quad \dot{y} = -\frac{g}{L}\sin x = -\frac{g}{L}x + \frac{g}{L}(x - \sin x)$$

and

$$\frac{|x - \sin x|}{|x| + |y|} \leq \frac{|x - (x - \frac{x^3}{3!} + \cdots)|}{|x|} \to 0 \quad \text{as} \quad |x| + |y| \to 0.$$

d) Not all nonlinearities satisfying the above limit need to be smooth polynomials. The second order equation $\ddot{x} + 2\dot{x} + x + x^{4/3} = 0$ becomes the system

$$\dot{x} = y, \quad \dot{y} = -x - 2y - x^{4/3},$$

and

$$\frac{\|f(\underset{\sim}{x})\|}{\|\underset{\sim}{x}\|} \leq \frac{|x^{4/3}|}{|x|} = |x^{1/3}| \to 0 \quad \text{as} \quad |x| + |y| \to 0.$$

Remark. This is an opportunity to point out the handy "little o" notation: $\|f(\underset{\sim}{x})\| = o(\|\underset{\sim}{x}\|)$ as $\|\underset{\sim}{x}\| \to 0$ means $\|f(\underset{\sim}{x})\|/\|\underset{\sim}{x}\| \to 0$ as $\|\underset{\sim}{x}\| \to 0$.

Note that $\|f(\underset{\sim}{x})\| = o(\|\underset{\sim}{x}\|)$ as $\|\underset{\sim}{x}\| \to 0$ implies $f(\underset{\sim}{0}) = \underset{\sim}{0}$ which means $\underset{\sim}{x}(t) \equiv \underset{\sim}{0}$ is a solution of the system $\dot{\underset{\sim}{x}} = A\underset{\sim}{x} + f(\underset{\sim}{x})$, or 0 is an equilibrium point of the system. Under these more general growth conditions on $f(\underset{\sim}{x})$ we would expect that if the solution $\underset{\sim}{x}(t) \equiv \underset{\sim}{0}$ (or equivalently the equilibrium point $\underset{\sim}{x}_0 = \underset{\sim}{0}$) of the linearized system is asymptotically stable, then so is the solution $\underset{\sim}{x}(t) \equiv 0$ of the nonlinear (almost linear) system.

We sketch a proof of this conjecture which seems a nice way to end this ODE sojourn. First of all, if $\Phi(t)$ is the fundamental matrix associated with the system $\dot{\underset{\sim}{x}} = A\underset{\sim}{x}$, $\Phi(0) = I$, then we can employ the same trick used in the proof of the previous stability result. If $\underset{\sim}{x}(t)$ is the solution satisfying the I.C. $\underset{\sim}{x}(0) = \underset{\sim}{x}_0$ then it satisfies the integral equation

$$\underset{\sim}{x}(t) = \Phi(t)\underset{\sim}{x}_0 + \int_0^t \Phi(t-s)f(\underset{\sim}{x}(s))ds.$$

The expression is ripe for Gronwall-ing but we first need some assumptions and estimates.

Assume the eigenvalues of A all have negative real parts, so there exist positive constants R, α such that $\|\Phi(t)\| \le Re^{-\alpha t}$, $t \ge 0$. Then

$$\|\underset{\sim}{x}(t)\| \le Re^{-\alpha t}\|\underset{\sim}{x}_0\| + \int_0^t Re^{-\alpha(t-s)}\|f(\underset{\sim}{x}(s))\|ds$$

or

$$\|\underset{\sim}{x}(t)\|e^{\alpha t} \le R\|\underset{\sim}{x}_0\| + \int_0^t Re^{\alpha s}\|f(\underset{\sim}{x}(s))\|ds.$$

Next note that the growth condition $\|f(\underset{\sim}{x})\| = o(\|\underset{\sim}{x}\|)$ implies that given any $\epsilon > 0$ (the fearsome ϵ at last!) there exists a $\delta > 0$ (and its daunting side kick!) such that $\|f(\underset{\sim}{x})\| < \epsilon\|\underset{\sim}{x}\|$ if $\|\underset{\sim}{x}\| < \delta$.

Since $\underset{\sim}{x}(t)$ is assumed to be close to the zero solution, we can let $\|\underset{\sim}{x}_0\| < \delta$. But $\underset{\sim}{x}(t)$ is continuous, so $\|\underset{\sim}{x}(t)\| < \delta$ for some interval $0 \le t \le T$, and the great Gronwall moment has arrived! We substitute $\epsilon\|\underset{\sim}{x}(s)\|$ for $\|f(\underset{\sim}{x}(s))\|$ in the previous estimate, Gronwall-ize, and obtain

$$\|\underset{\sim}{x}(t)\|e^{\alpha t} \le R\|\underset{\sim}{x}_0\| \exp\left[\int_0^t \epsilon R\, dt\right], \qquad 0 \le t < T,$$

or

$$\|\underset{\sim}{x}(t)\| \le R\|\underset{\sim}{x}_0\|e^{(\epsilon R - \alpha)t}, \qquad 0 \le t < T,$$

the crucial estimate.

Since $\epsilon > 0$ is at our disposal and so is x_0, let $\epsilon < \alpha/R$ and choose $\underset{\sim}{x}_0$ satisfying $\|\underset{\sim}{x}_0\| < \delta/2R$. The estimate gives us that $\|\underset{\sim}{x}(t)\| < \delta/2$ for $0 \le t < T$, and we can apply a continuation argument interval by interval to infer that

i) Given any solution $\underset{\sim}{x}(t)$, satisfying $\underset{\sim}{x}(0) = \underset{\sim}{x}_0$ where $\|\underset{\sim}{x}_0\| < \delta/2R$, it is defined for all $t \ge 0$ and satisfies $\|\underset{\sim}{x}(t)\| < \delta/2$. Since δ can be made as small as desired this means the solution $\underset{\sim}{x}(t) \equiv 0$ (or the equilibrium point $\underset{\sim}{0}$) is *stable*.

ii) Since $\epsilon R - \alpha < 0$ this implies that $\lim_{t \to \infty} \|\underset{\sim}{x}(t)\| = 0$ hence $\underset{\sim}{x}(t) \equiv 0$ is *asymptotically stable*.

We have proved a version of the cornerstone theorem, due to Perron and Poincaré, for almost linear systems:

Given $\dot{\underset{\sim}{x}} = A\underset{\sim}{x} + \underset{\sim}{f}(\underset{\sim}{x})$ where $\underset{\sim}{f}$ is continuous for $\|\underset{\sim}{x}\| < a$, $a > 0$, and $\|\underset{\sim}{f}(\underset{\sim}{x})\| = o(\|\underset{\sim}{x}\|)$ as $\|\underset{\sim}{x}\| \to 0$. If all the eigenvalues of A have negative real parts then the solution $\underset{\sim}{x}(t) \equiv \underset{\sim}{0}$ is asymptotically stable.

Remark.

i) The result can be generalized to the case where $\underset{\sim}{f} = \underset{\sim}{f}(t, \underset{\sim}{x})$, is assumed to be continuous for $\|\underset{\sim}{x}\| < a$, $0 \le t < \infty$, and the estimate $\|\underset{\sim}{f}(t, \underset{\sim}{x})\| = o(\|\underset{\sim}{x}\|)$ as $\|\underset{\sim}{x}\| \to 0$ is uniform in t. For example, $|x^2 \cos t| = o(|x|)$ as $|x| \to 0$, uniformly in t, since $|x^2 \cos t| \le |x|^2$.

ii) By considering the case $t \to -\infty$ and assuming all the eigenvalues of A have positive real parts, the proof above implies that if $\|\underset{\sim}{f}(\underset{\sim}{x})\| = o(\|\underset{\sim}{x}\|)$ as $\|\underset{\sim}{x}\| \to 0$ then the solution $\underset{\sim}{x}(t) \equiv 0$ is unstable.

iii) For the two-dimensional case it follows that if the origin is a stable (unstable) spiral for the linear system, it is a stable (unstable) spiral for the almost linear system, and similarly for stable (unstable) nodes. It can be shown that the saddle point configuration is preserved, whereas a center may remain a center or be transformed into a stable or unstable spiral.

The Perron/Poincaré theorem gives credence to the assertion that when we linearize around an equilibrium point $\underset{\sim}{x}_0$ by using Taylor's Theorem, the asymptotic stability or instability of the linear system is preserved. The theorem is a fitting milestone at which to end this brief, eclectic journey.

FINALE

To those who have snarled and staggered this far, the author hopes you have been stimulated by this small tome, and will continually rethink your approach to the subject. And ALWAYS REMEMBER

The subject is

> ORDINARY DIFFERENTIAL EQUATIONS
> and NOT
>> Algebra
>> Calculus
>> Linear Algebra
>> Numerical Analysis
>> Computer Science

References

(And other tracts the author has enjoyed, with brief commentary.)

D. Acheson, *From Calculus to Chaos, An Introduction to Dynamics,* Oxford University Press, Oxford, 1997.

> A delightful book to waltz through the calculus, and the final chapters are a nice introduction to ODEs and chaotic systems.

W.E. Boyce and R.C. DiPrima, *Elementary Differential Equations and Boundary Value Problems,* 6th Ed. Wiley, New York, 1997.

> Old timers will remember the thin, classic grandfathers of this book, which has now grown plump like its contemporaries. But it is still an excellent introductory text.

F. Brauer and J.A. Nohel, *The Qualitative Theory of Ordinary Differential Equations, An Introduction,* W.A. Benjamin, 1969; (reprinted) Dover Publications, New York, 1989.

> One of the first textbooks on the modern theory; the discussion of Lyapunov theory is excellent.

F. Brauer and D.A. Sánchez, Constant rate population harvesting: equilibrium and stability, *Theor. Population Biology* **8** (1975), 12–30.

> Probably the first theoretical paper (but readable) to discuss the effects of harvesting on equations of population growth. Sandhill Cranes are featured.

M. Braun, *Differential Equations and Their Applications: An Introduction to Applied Mathematics,* 4th Ed. Springer-Verlag, New York, 1992.

> A well written introductory textbook with a solid emphasis on applications. The discussion of art forgery and Carbon-14 dating is unsurpassed.

R.L. Burden and J.D. Faires, *Numerical Analysis,* 5th Ed. PWS-Kent, Boston, 1993.

> An easy to read introduction to numerical methods with wide coverage and good examples.

G.F. Carrier and C.E. Pearson, *Ordinary Differential Equations,* Blaisdell, 1968; (reprinted) Soc. for Industrial and Applied Mathematics, Philadelphia, 1991.

> Snapshot-like coverage of a lot of classic topics important in applications, with problems that are formidable.

F. Diacu, *An Introduction to Differential Equations, Order and Chaos,* W.H. Freeman, New York, 2000.

> A recent text which bucks the current trend towards obesity, and has a very modern flavor. If you believe an introductory book should cover 1-1 1/2 semesters of topics, like the original Boyce and DiPrima text, look at this one.

K.O. Friedrichs, *Advanced Ordinary Differential Equations,* Gordon and Breach, New York, 1966.

> One of the volumes of the old Courant Institute Notes which were all great contributions to modern mathematics.

W. Hurewicz, *Lectures on Ordinary Differential Equations,* MIT Press, Cambridge, 1958.

> Truly a classic! This little paperback was one of the first books to introduce the modern theory, as developed by the Soviet school, to English-speaking audiences.

D.W. Jordan and P. Smith, *Nonlinear Ordinary Differential Equations,* Oxford University Press, Oxford, 1987.

> Lots of wonderful examples of nonlinear systems and the techniques needed to study their behavior—a very handy book to have on the shelf.

E. Kamke, *Differentialgleichungen, Losungsmethoden und Losungen,* Chelsea, New York, 1948.

> If it can be solved, no matter how weird it looks, it's probably in here! A more recent example of the same mad pursuit is a handbook by A.D. Polyanin and V.F. Zaitsev (CRC Press, 1995) which has over 5000 ODEs and their solutions, including 267 Riccati equations!

W.D. Lakin and D.A. Sánchez, *Topics in Ordinary Differential Equations,* Prindle, Weber and Schmidt, 1970; (reprinted) Dover Publications, New York, 1982.

> Covers asymptotics, singular perturbation, nonlinear oscillations, and the relation between boundary value problems and elementary calculus of variations.

A.C. Lazer and D.A. Sánchez, Periodic equilibria under periodic harvesting, *Math. Magazine* **57** (1984), 156–158.

> The authors thought this would be a nice paper for a student journal, consequently, its useful major result is almost unknown to practicing mathematical ecologists.

J.E. Leonard, The matrix exponential, *SIAM Review* **38** (1996), 507–512.

> If you really want to know how to compute $\exp(tA)$ here's a good way to do it.

L.N. Long and H. Weiss, The velocity dependence of aerodynamic drag: a primer for mathematicians, *American Mathematical Monthly* **106** (1999), 127–135.

> Imparts some wisdom into the oft stated and misused assumption that the air resistance of a falling body is proportional to the square of its velocity.

R.E. O'Malley, Jr., *Thinking About Ordinary Differential Equations,* Cambridge University Press, Cambridge, 1997.

> A small paperback which is an excellent introduction, with fine problems, and of course, O'Malley's favorite topic, singular perturbations, gets major billing.

R.E. O'Malley, Jr., Give your ODEs a singular perturbation!, *Jour. Math. Analysis and Applications* **25** (2000), 433–450.

> If you want a flavor of singular perturbations, this paper is a very nice collection of examples, starting with the simplest ones.

K.R. Meyer, Jacobi elliptic functions from a dynamical systems point of view, *American Mathematical Monthly* **108** (2001), 729–737.

> The solutions and periods of conservative systems are intimately connected to elliptic functions and elliptic integrals. The author gives a first-rate introduction to this connection.

V. Pliss, *Nonlocal Problems of the Theory of Oscillations,* Academic Press, New York, 1966.

> A leading Soviet theorist gives a thorough introduction to the subject. It is an older treatment but is one of the few books that discusses polynomial equations.

J. Polking, A. Boggess, and D. Arnold, *Differential Equations,* Prentice Hall, Upper Saddle River, 2001.

> A recent big, introductory book, with a nice interplay between computing and theory, plus a very good discussion of population models.

D.A. Sánchez, *Ordinary Differential Equations and Stability Theory: An Introduction,* W.H. Freeman, 1968; (reprinted) Dover Publications, New York, 1979.

> The author's attempt, when he was a newly-minted PhD to write a compact introduction to the modern theory, suitable for a graduate student considering doing ODE research, or for an undergraduate honors course. Its price may help maintain its popularity.

D.A. Sánchez, Ordinary differential equations texts, *American Mathematical Monthly* **105** (1998), 377–383.

> A somewhat acerbic review of the topics in a collection of current leviathan ODE texts. The review played a large role in the inspiration to write the current book.

D.A. Sánchez, R.C. Allen, Jr., and W.T. Kyner, *Differential Equations,* Addison-Wesley, Reading, 1988.

> Yes, this was probably one of the first corpulent introductory texts, now interred, but made the important point that numerical methods could be effectively intertwined with the discussion of the standard types of equations, and not as a stand-alone topic. R.I.P.

S.H. Strogatz, *Nonlinear Dynamics and Chaos,* Addison-Wesley, Reading, 1994.

> Gives a very thorough introduction to the dynamical systems approach, with lots of good examples and problems.

S.P. Thompson, *Calculus Made Easy,* Martin Gardner (ed.), originally published 1910; St. Martin's Press, New York, 1998.

> This reference is here because in the 1970's the author wrote a letter in the *Notices of the American Mathematical Society* protesting the emergence of telephone directory size calculus books. The creation of the Sylvanus P. Thompson Society, honoring the author of this delightful, incisive, brief tome, was suggested to encourage the writing of slim calculus texts. Perhaps a similar organization is needed today, considering the size of ODE books.

F. Verlhulst, *Nonlinear Differential Equations and Dynamical Systems,* Springer-Verlag, Berlin, 1996.

> A very well written and thorough advanced textbook which bridges the gap between a first course and modern research literature in ODEs. The mix of qualitative and quantitative theory is excellent with good problems and examples.

Index